Junkers Ju 287

Junkers Ju 287

The World's First Swept-Wing Jet Aircraft

Horst Lommel

Schiffer Military History
Atglen, PA

Translated from the German by Don Cox
Book Design by Ian Robertson.

Library of Congress Catalog Number: 2004102325

Printed in China.
ISBN: 0-7643-2059-9

This book was originally printed under the title,
Junkers Ju 287 Der erste Jet-Bomber der Welt und weitere Pfeilflügelprojekte
by Aviatic Verlag

We are interested in hearing from authors with book ideas on related topics.

Published by Schiffer Publishing Ltd.
4880 Lower Valley Road
Atglen, PA 19310
Phone: (610) 593-1777
FAX: (610) 593-2002
E-mail: Info@schifferbooks.com.
Visit our web site at: www.schifferbooks.com
Please write for a free catalog.
This book may be purchased from the publisher.
Please include $3.95 postage.
Try your bookstore first.

In Europe, Schiffer books are distributed by:
Bushwood Books
6 Marksbury Avenue
Kew Gardens
Surrey TW9 4JF
England
Phone: 44 (0) 20 8392-8585
FAX: 44 (0) 20 8392-9876
E-mail: Bushwd@aol.com.
Free postage in the UK. Europe: air mail at cost.
Try your bookstore first.

Contents

Foreword

Toward the end of 1999 the military institutions of the Western World were jarred from their slumber by some unexpected news. The Russian Air Force, supposedly on its last legs, presented a new and unusual combat jet surrounded in secrecy to an astounded public - the Sukhoi Su-37 *Berkut* (Bald Eagle). This was a fighter design of a special kind. The primary lifting segments were two negatively swept wings followed by a delta-shaped empennage and, what was most amazing, the *Berkut* utilized lift-inducing canards attached to the forward fuselage section.

This distinctive design, to which was credited a superior maneuverability in dogfighting as well as exceptional low-speed handling qualities (suited, for example, to carrier operations), was revealed unannounced to an astonished crowd as part of an air show line-up.

The above-mentioned incident provides a good opportunity to mention an equivalent development that planted the seeds for the development of futuristic technologies some fifty years earlier. We are only just now able to begin harvesting the fruits, thanks to the availability of advanced materials.

At the time, the Junkers Works in Dessau began working on a developmental line shortly before the end of the war. Their goal was to design and build a heavy jet bomber, a STRABO (from *Strahlbomber* - jet bomber) for short, with swept wings that would be able to reach near-sonic speeds. The application of a negative wing sweep was the distinguishing trademark of what would become the Junkers Ju 287 six-engined design.

This book is an attempt to detail the historic events and facts pertaining to this jet bomber and to correct any inaccuracies and omissions in the few publications on the type that have appeared thus far. Using surviving meeting reports and conference minutes, some of which are reproduced here in their entirety and offering the reader a "you were there" feeling, virtually the entire developmental course of the Ju 287 can be followed with all its ups and downs. Personal experiences from contemporary eyewitnesses add a bit of spice to the story.

Inseparably linked with the saga of the technologically interesting Junkers Ju 287 is the developmental history of the Junkers EF 131. This was effectively a resurrected Ju 287, begun in Germany and completed in the Soviet Union.

Anyone who writes about the Junkers Ju 287 can't avoid mentioning the Heinkel He 343, a design that was drawn up out of the necessity of offering a competing design. Constant bickering and power struggles, as well as jealous quarreling between the two companies run like a red thread through the developmental history of the jet bomber, even though the two types were not actually competing head-to-head as they represented two entirely different performance classes.

In addition to this book's main theme - the history of the Junkers Ju 287 - other swept-wing projects are examined. These, however, have been selected to flesh out the history of swept wing development and are limited to those projects that either actually flew or those that were directly competing with the Ju 287. Several of the swept-wing projects conceived for supersonic flight were indeed tested in Germany, the USA or the former Soviet Union and formed the basis of today's supersonic jets, swing-wing aircraft, etc.

Two aircraft types are repeatedly mentioned as competitors of the Ju 287, the Messerschmitt Me P 1107/1108, which only gets an honorable mention because of the theory behind it, and the Horten Ho XVIII that, despite its unconventional flying wing layout, was awarded a manufacturing contract just prior to the end of the war.

The four remaining swept-wing projects that are examined in closer detail within the pages of this book are the DFS 346, the Lippisch Li P 13a (DM 1), the Horten Ho XIII (Ho X), and the Messerschmitt P 1101. Amazingly, these were all designed to break through the hitherto impenetrable sound barrier. These four supersonic projects, all of which advanced to the prototype stage, fascinate us because of the different approaches they represent in their designers' attempts at conquering the sound barrier.

Naturally, there's something here for the model builder. I'd like to take just a moment and note something on behalf of the model builder. It is this group of persistent enthusiasts who constantly notice easily overlooked details and point them out to the likes of us. What good is it to have the most beautiful and unique photographs if we don't recognize what interesting details are hidden there.

I am conscious of the fact that this book tackles a difficult subject, for - as mentioned earlier - on the one hand the available documentation is sparse and can only be thus, given that very few test flights were completed before the war's end and the type never reached production readiness. On the other hand there have been many incongruities caused by contradicting statements from eyewitnesses and such-and-such an expert (or those who claim to be). Furthermore, there are those who possess unique material and want to continue being the only ones to possess it. As a rule, these people then write in after a successful publication to denounce supposed omissions, but that is another story.

All the more, then, that this book is also a call to those eyewitnesses still living to let themselves be known and share with us their experiences with this interesting design, so that they can be considered for any future edition.

Nor have we forgotten to include the Junkers Ju 322 in the theme of swept-wing designs. This is considered an absolute failure and often goes unrecognized or treated in a superfluous manner, but we've given it a complete chapter here. In this context, there is something of a drop of bitterness that needs mentioning. The actual original photographs of the Ju 322 reproduced in these pages come from a private source. They were taken surreptitiously with amateur cameras while driving by and are therefore not of particularly outstanding quality, but are unique nonetheless.

As well as a few passages reserved for the person of Hans Kammler, there is also the mention of the DFS 346 as well as the subject of rocket assisted takeoff engines. I've already written much about these in other publications, but as they are indispensable to any complete discussion of swept wings they are of necessity included in these pages as well.

Before I leave you, dear reader, to peruse what I hope you will find a not only technically interesting but an also exciting read, I would like to send you on your way with a fitting quote by Professor Ernst Heinkel:

> "To be truly successful and give new developments room to breathe, one must risk crossing the threshold."

Siegen, December 2002
Horst Lommel

The Road to Negative Wing Sweep

The development of the first jet engines and the functionally reliable liquid-fueled rocket engines in the late 1930s led to the ever increasing probability that it would soon be possible to attain transonic speeds, and even break through the hitherto unconquerable sound barrier itself. *Flugkapitän* Heini Dittmar was one of the first pilots to experience what happens when a plane approaches the speed of sound when he reached a speed of 1,003.67 km/h during a test flight in the Me 163 (a liquid-fueled rocket-powered fighter interceptor) on 2 October 1941. Heini Dittmar later provided the following information about his experience on this historic flight:

Everything was ready for the 2nd of October. The route for the test run had been precisely laid out and the cinematic theodolites (equipment used to measure flight path and provide speed analysis) were set up when the Me 110 took off with my fully fueled Me 163

Test pilot Heini Dittmar, in white overalls, preparing for a flight. Pre-flight checks for a test flight with the rocket-powered DFS 194.

in tow. I released the tow cable at an altitude of about 4,000 meters, nosed the plane over a bit into a shallow glide to check everything out once again, then lit the engine. The speed indicator gauge began to race upward as soon as I applied full thrust: 500-600-700-800-900 km/hr. And by the time I'd reached the test run area the instrument needle was quivering at just over the 1,000 km/h mark. But before I could really savor the fact that I'd crossed this dream threshold I felt the stabilizer begin to vibrate violently. The speedometer needle started dancing wildly, the plane began to porpoise crazily and then bunted over into an uncontrollable dive. Nothing I did could get the plane back under control. I thought I was a goner, but then I succeeded in wresting back control at an altitude just under 2,000 meters. It wasn't until then that I noticed that I sat in the machine bathed in sweat. Face, back, hands - everything was dripping wet, so that my goggles fogged and I couldn't see out. I took the goggles off and carefully cleaned the inner lenses with my fingers. Once I could more or less see again I was able to land the aircraft without any further problems.

In his near-supersonic flight, Heini Dittmar was the first person in the world to experience this unpleasant phenomenon. Efforts were now devoted to find solutions to eliminate such problems for transonic flights in the future as a step toward true supersonic flight.

In order to put the obstacles to this endeavor into proper perspective it is necessary for us to make a short detour into the theory of supersonic flight.

It is a general given that the lifting forces of a wing profile, dependent upon airflow velocity, are the determinants for the flight handling qualities of an aircraft.

Let's imagine for a moment that a wing model is mounted inside a wind tunnel where the airflow velocity over the control surface is increased at a constant rate. For speeds under the sound barrier the wing profile remains undisturbed by the air molecules flowing around it. We will use the letter "v" to represent the flow velocity and the letter "c" for the speed of sound. This results in the equation v/c=M, whereby M equals the so-called Mach number. The picture changes dramatically as the sound barrier is approached. The air molecules can no longer flow off fast enough; in effect they compress themselves ahead of the wing leading edge. This in turn results in airflow over the wing surface with insufficient velocity to create adequate lift. The complexities involved with pressure ratio, particularly as it occurs on the upper wing surface, are not addressed here as we don't want to become too bogged down in theoretical discussion. Suffice it to say that the air molecules no longer flow over the control surface profile. This leads to turbulence and the airflow over the wing, the boundary layer, now separates. This separation and the accompanying increase in drag as the speed of sound is approached are the main problems facing the mechanics of flight when penetrating the sound barrier. This phenomenon, which we've only discussed in a very simplified manner, also affects an airplane's flight handling characteristics in a radical way. Most airplanes that are not designed for penetrating

Photograph of a combat-ready Me 163B-1 *Komet* as flown operationally in large numbers with JG 400.

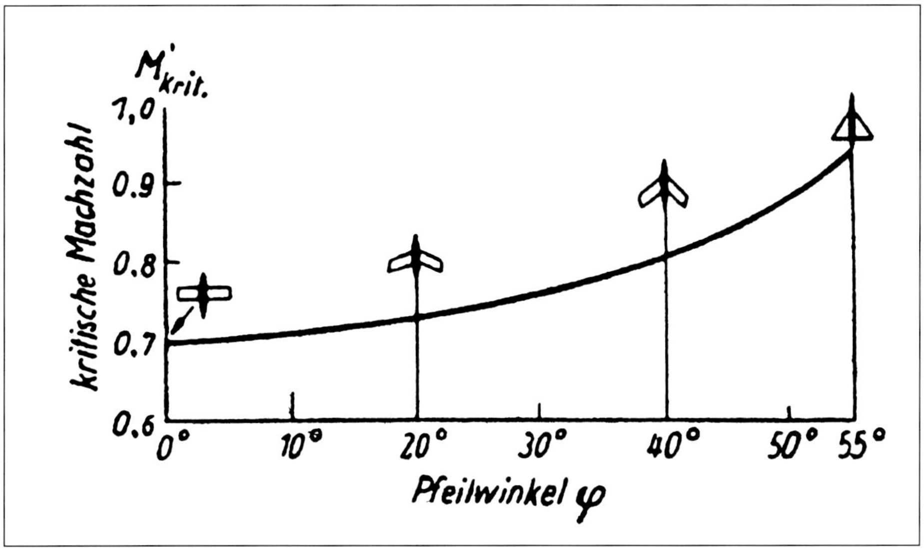

The Y coordinates on the curve chart shows the critical Mach number *M* in relation to the wing sweep angle. It is quite obvious that a greater sweep angle (*Pfeilwinkel*) offers a greater potential critical Mach number (*Machzahl*).

the sound barrier become nose heavy as the speed approaches that of sound and will nose over into a dive (as Heini Dittmar discovered on his test flight). When breaking the sound barrier the compressed molecules of air suddenly break up, produce the well-known sonic boom, and then resume their uninterrupted flow along the profile of the control surface. Premature supersonic flow can occur along some localized areas of the wing, which can suddenly cause the pressure ratio to change on the upper wing surface. This phenomenon is known as a shock wave. Such a shock wave can lead to the airflow separating at one point on the control surface before another. This event has cost the lives of many test pilots, as it ushers in such a strong roll moment that the pilot seldom has a chance to recover the aircraft.

What, then, can be done to avoid all this?

The first thing noticed was that with a straight wing the drag produced by the compressed air molecules built up at a virtually constant rate along the entire length of the wingspan. On the other hand, a so-called swept wing produced significantly different results, as it did not possess a straight wing centerline but rather one that angled to the rear. This type of wing design was called a positively swept wing. The larger the angle, the less drag is built up.

The swept wing, therefore, is a critical feature when successfully designing an airplane for supersonic flight, but is not the sole determining element. In addition, there is the shape of the profile itself. The thinner the wing's profile, the better it is for attaining supersonic flight.

Now, one might be forgiven for assuming that these are all the ingredients needed for a successful aircraft design for transonic and supersonic operations. According to this reasoning, all you would need is a thin wing profile paired with a sharply swept wing mated to a suitable fuselage, and you would then have your desired results.

The design of the DFS 346 reveals that this is not the case. The DFS 346 was an supersonic research plane that the Siebel Works began building in Halle before the end of the war, but was finished and tested on Russian soil after the war's end. The development of the DFS 346 is a good example of the effects of inadequate design resulting from exploring completely new territory.

Since the DFS 346 was expected to travel at hitherto unattainable speeds at altitudes up to 30,000 meters, its pressurized cockpit was fitted with an elongated, pointed nose favorable to airflow. This appendage made use of plexiglass glazing. The pitot tube for measuring velocity was also affixed to the cockpit nose. The overall fuselage had a circular cross-section; the profile resembled that of a rocket body with the forward section corresponding to the NACA profile 00121-0.66-50.

The designers entered completely uncharted territory with the layout of the wing. Never before had a wing been tested in the air with such an extreme sweep of 45 degrees. The only information they had was from innumerable wind tunnel tests that had been carried out in the late summer of 1944 using models built by DFS.

These test results were reflected in the layout and position of the ailerons and the shape of the empennage. The selection of a symmetrical NACA 0.012-0.55-1.25 wing profile was another result of the tests. An additional unique feature of the wings was the use of separate split ailerons that, when deployed would extend at different angles. The T-tail's elevators, also in two sections were designed on the same principle. The designers hoped that these features would offer improved stability at the anticipated higher speeds.

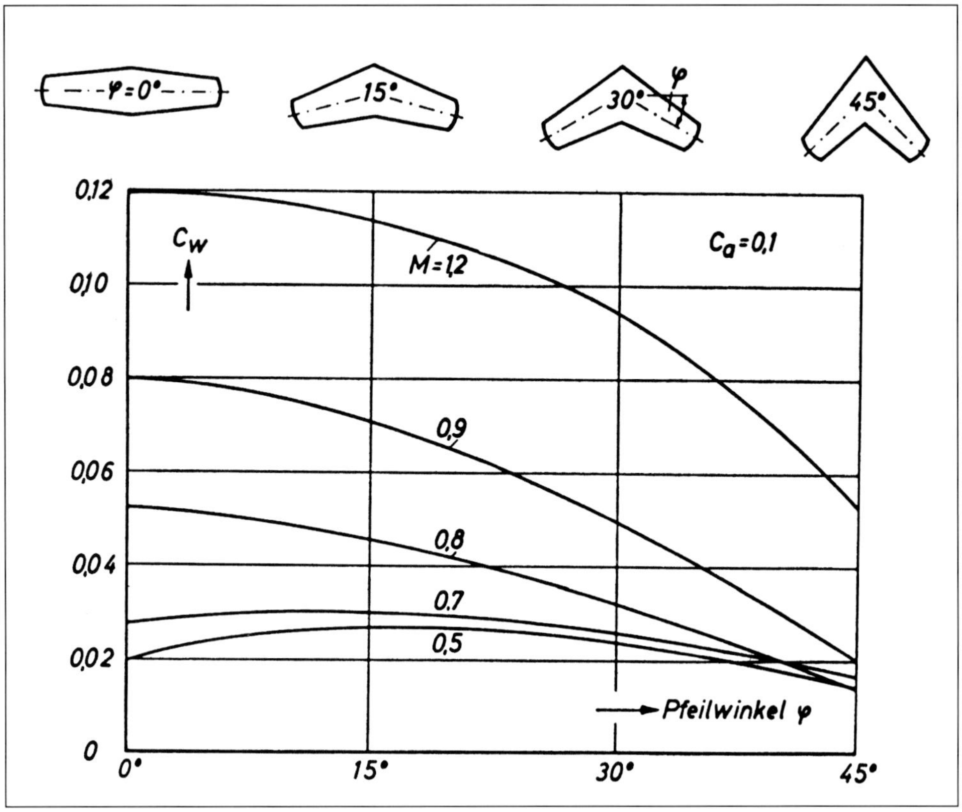

The diagram shows the effect of the sweep angle on the drag coefficient c_w in the Mach number range of $M = 0.5$ - $M = 1.2$, with a lift coefficient of $c_a = 0.1$.

There was considerable disagreement over the stress to be expected when penetrating the sound barrier. Since, however, the assumption was that the stress would be immense, the entire aircraft was of metal construction. Any engine other than rocket propulsion was out of the question. All participants understood that the engine needed to accelerate the DFS 346 to supersonic speeds did not exist at that time. For this reason the aft fuselage section was designed generously enough to accommodate the installation of two Walter HWK 109-509 rocket motors. The combined power of the two engines would theoretically be able to heave the DFS 346 from its release altitude of 7,500 meters to over 30,000 meters and there accelerate it to over 2,000 km/h. Following a successful flight, it would then glide back to its starting point and land on an extendable skid.

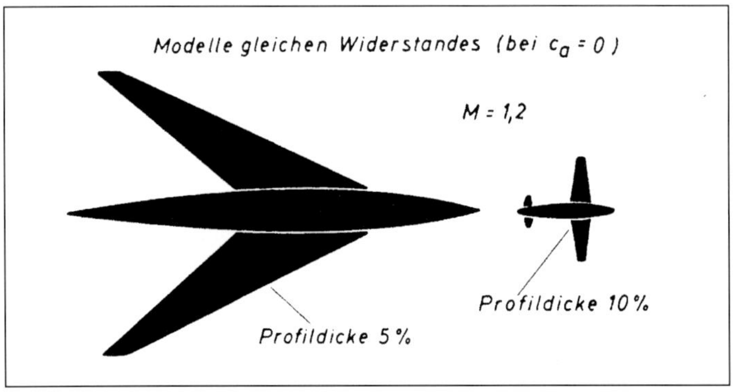

Schematic drawing of two aircraft in the same scale. It shows that, thanks to its swept wing layout and wing chord, the much larger left aircraft has the same drag at Mach 1.2 as the right one with its conventional design (*Profildicke* = wing thickness).

Plans first called for the construction of an unpowered prototype in order to assess the basic characteristics of the swept wing and the flight handling of the aircraft at different subsonic speeds. Then, a single Walter rocket engine would either be fitted to this airplane or another prototype with the goal of testing the behavior of the aircraft in rocket-powered subsonic flight. Not until after these tests would two Walter rocket motors be installed in a test plane, which would gradually fly at increased speeds in stages until the sound barrier "adventure" was complete. As early as the autumn of 1944, the design blueprints had advanced to the point where the first construction drawings could be turned over to the Siebel *Flugzeugwerke* in Halle. Since DFS had little experience with metal aircraft construction, the construction and tweaking of the design plans were handed over to Siebel.

Siebel immediately began working on the first DFS 346 example. First, a full-scale wooden mockup was built, reputedly completed shortly before the end of the war. The first flying, non-powered version, designated the DFS 301, also was reputed to have been in the advanced stages of construction when the war ended. Additional prototypes or subassemblies are rumored to have been under construction when the Siebel Works were overrun by the

Russian Army. Nevertheless, there is much conflict of opinion on the state of completion at the end of the war. Be that as it may, the fact is the technical officers of the Soviet Army considered the DFS 346 to be so interesting and important that, after a detailed examination, the decision was made to consider all participants on the project - whom they had the right to apprehend - as guests of the Soviet Union and to deport them along with their family members back to Mother Russia. It's worth mentioning an interesting side note to the history of the DFS 346 at the war's end. In his book "*War Prizes*", Phil Buttler writes on page 50 that the Americans had a wish list of booty in which the DFS 346 was at the number 1 position, ahead of the Bachem Ba 349, and that the Americans were the first to acquire the DFS 346 before later handing it over to the Russians when they occupied the area around Halle.

All assembly equipment for the DFS 346 was dutifully taken apart at the Siebel Works in Halle and taken - together with the prototypes in their various stages of construction - to the Soviet Union. Transfer operations had been completed by the summer of 1946. The design offices were reestablished some 120 km from Moscow, near the city of Dubna, outside the tiny village of Podberesye. Design work was to continue there under the watchful eyes of the Soviets.

The Germans were allowed to move freely without restrictions inside their "Construction Camp". The design department, now designated OKB 2, operated under the direction of Hans Rössling, who had formerly been responsible for managing the DFS 346 project with Siebel. In addition to the equipment and partially completed aircraft prototypes, the project designation also came to Russia - with one restriction. No longer was it known as the DFS 346, but simply and succinctly just "346".

The engine-less prototype brought from Halle was completed by the fall of 1946. It had not been intended for flight testing and was instead tested in the ZAGI's (Russian Research Institute) wind tunnel T-101. There it was discovered that at higher angles of attack there was a rather pronounced separation of the boundary layer near the wingtips that quickly spread over the entire wing surface and, for all intents and purposes, destroyed any longitudinal stability the DFS 346 might have had. It now became apparent that, because of ignorance and lack of experience - this was completely uncharted territory, after all - there had been a major error in calculating the wing layout. The decision to use the same profile throughout the length of the wing had turned out to be a poor one. What was amazing, though, was the fact that even with a total separation of the boundary layer the ailerons remained fully effective. Wind tunnel testing also revealed other problems, although these were not as critical.

In order to deal with the greater problem of the rapid separation of airflow, so-called boundary layer fences were affixed to the upper wing surface (see photo p. 15). This solution appears to have virtually corrected the problem. Boundary layer fences are flat plates conforming to the wing profile that are attached approximately two-thirds along the length of the wingspan. Their purpose is to prevent the boundary layer from drifting toward the tips of the wings. This type of boundary layer fence for positive sweepback is based on a discovery by DVL aerodynamicist Dr. Wolfgang Liebe. Once wind tunnel testing showed that the most critical defects had been cor-

Wind tunnel model of the DFS 346.

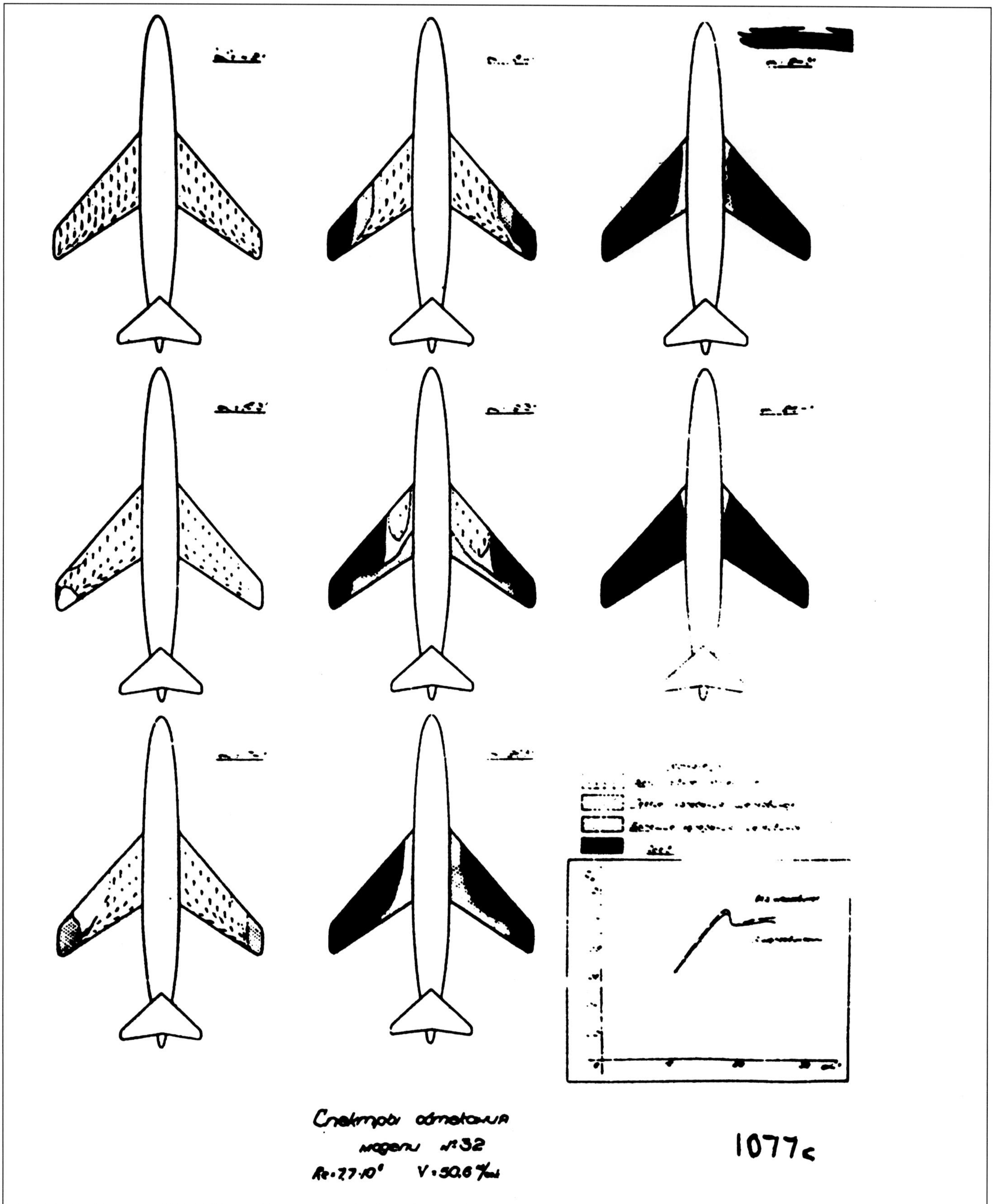

Results of wind tunnel testing. The sequence clearly shows the course of airflow separation, beginning at the wing tips and continuing until separation is complete.

rected, the improvements were applied to the prototypes then under construction. By this time work had finished on the unpowered glider version, the "346 P". The aircraft was transported to the airfield at Tyopli Stan, a restricted airbase near Moscow, in early 1948; it was there that the first test flights were carried out.

The developmental history of the DFS 346 clearly shows us that there were not just advantages bundled with the swept wing, but several disadvantages as well. These included separation of the airflow and resulting dangerous instability along the longitudinal axis as well as poor low-speed handling, to name just the two most important ones.

At this point it is necessary to look at theories a bit more. We've already learned about the Mach number M. As mentioned earlier, if air is blown over a curved profile at high subsonic speeds, the air molecules on the upper surface accelerate at a rapid rate. We call it the critical Mach number when the speed of sound is reached in local areas along the profile curve. Once the critical Mach number is passed there is a sharp increase in the drag coefficient, which in turn must be offset by a corresponding increase in thrust. Which posed a problem for the first generation of jet engines still in their infancy. There was only one solution to overcoming this state of affairs: reducing the critical Mach number by increasing the sweepback. But this, too, is not the ultimate in wisdom, as we've learned from the DFS 346 fiasco. Thus was born a remarkably simple, yet brilliant idea. If you simply rotated the sweepback through 180 degrees, effectively mirroring the current layout, you would actually retain the positive characteristics with regard to transonic/supersonic flight, but expect that the boundary layer separation would take place much later on these wings - now with a sharp forward sweep, than was the case with the normal swept back wing configuration. Furthermore, you could expect significantly better low-speed handling. This wing design was called negative wing sweep, or forward sweep.

The 346 III rests on its landing skid after a successful test flight. Beneath the wings can be seen the auxiliary support skids, designed to prevent the aircraft from tipping over onto its wings. The photo also shows to good effect the boundary layer fences fitted to the wings.

Digging through old documents reveals that forward sweep played a role at a much earlier date, although not in connection with high speed flight. In the early 1920s a British sport plane was reputedly fitted with negatively swept wings. This aircraft was even logged by Landwerlin and Berreur for all eternity under patent number 196937. If we go further back, we come across swept wing designs by the Junkers Company as early as 1915-1918 that were tested in the in-house wind tunnel at Aachen and Dessau. Yes, you read it correctly - even as early as 1915 the Junkers Company had several wind tunnels for research purposes. The fact that the Junkers Company was already wrestling with swept wings in the gray dawn of aviation brings us back to the topic of forward sweep. This idea was to be the basis for an entirely new six-engine turbojet bomber that Junkers would develop. It wasn't too long, however, before one realized that this concept was not without its pitfalls. To be sure, the boundary layer flowed toward the fuselage as expected, leading the airflow to begin separating at the wing root and, in comparison with the positively swept wing, caused the plane to stall harmlessly. The problem, however, was with the fact that at the time there was no way to produce the requisite wing stiffness necessitated by forward sweep. This caused the wings to flex when turning, when pulling up, in gusty weather, etc., something that contributed to a reduction in longitudinal stability. However, this problem could be countered by a designer's trick; the wing was built with a twist in the opposite direction to that of the anticipated twisting forces, meaning that the longitudinal stability would only be minimally affected by the torque effects.

Another problem of forward sweep was the reduced directional stability, but this could be offset by using a correspondingly large rudder. Thus, one by one the major hurdles to realization of the project were overcome.

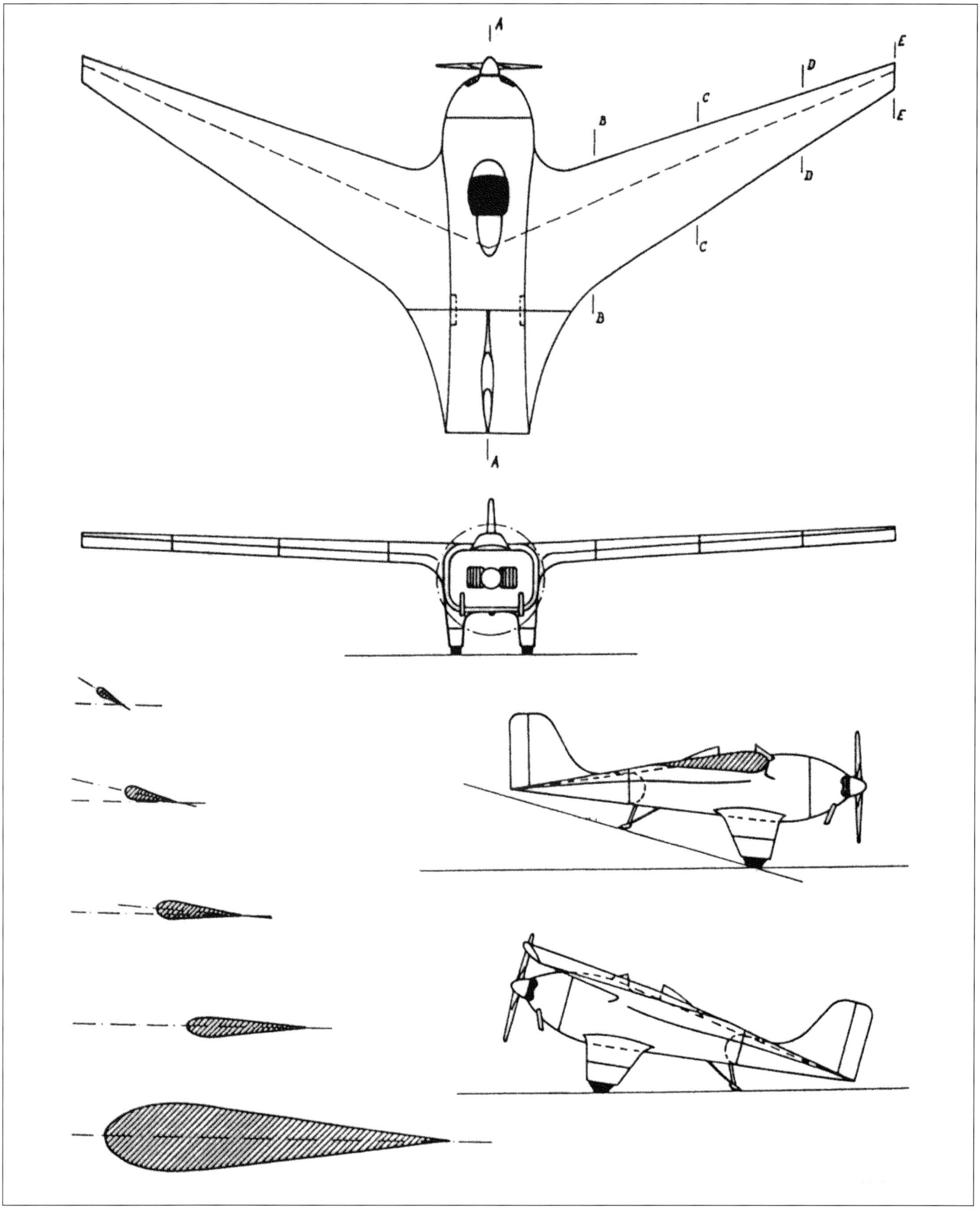

Four-view drawing of the British Landwerlin and Berreur design with negatively swept wings, from the early 1920s.

Operation STABO, the Junkers Ju 287 Jet Bomber

Professor Dr. Ing. Heinrich Hertel (Hertel had moved from the Heinkel Works to Junkers following his unfortunate involvement with the development of the Heinkel He 177) had always been a champion of sweepback for high speed flight. Early on, he'd studied the characteristics and nuances of every kind of swept wing in countless wind tunnel experiments. After these wind tunnel tests revealed certain advantages to forward sweep, the Junkers Company, on the recommendations of Professor Hertel, began working on the proposed six-engine jet bomber project with negative wing sweep in mind. The designers were well aware that the realization of such a project would be very time consuming, as they were breaking new ground in many respects. Under the direction of Dipl. Ing. Hans Wocke, the developmental team made the decision to first build an airplane to function as a test bed for researching the qualities of forward swept wings through experimental flights in order to confirm the overall positive wind tunnel results. Developmental work on the actual bomber design was to proceed parallel to the test plane.

A Heinkel He 177 in desperate need of repairs. The fuselages for the two Junkers Ju 287 testbeds came from such aircraft that no longer had any operational usefulness.

The Ju 288's cockpit evidenced a remarkable similarity to that of the Ju 287.

The result of this approach was a rather curious looking machine, that was more or less a kit plane put together from various components from a wide variety of existing operational aircraft. The fuselage came from a Heinkel He 177 A-3, the empennage and aft fuselage section from a Junkers Ju 188 G2 (according to Junkers documents. Other sources claim it was a Ju 388 L1), the main undercarriage wheels from the Ju 352 A1, and the twin-track nose gear from a captured B-24 Liberator bomber; the only newly designed component were the wings. From the outset, one of the recently available jet engines by Junkers, BMW or the Heinkel Works was planned as the powerplant in order to reach the higher flight speeds. The reasoning behind the idea of building a test platform

The Junkers Ju 287 V1 prior to its maiden flight at Brandis. The checkerboard pattern of the thermopaint is clearly recognizable behind the fuselage engine pod. This paint was designed to register the temperature of the outer fuselage skin. (photo: *Deutsches Museum*)

using components from existing aircraft was primarily the intent to create a test bed in the shortest amount of time possible and thereby have test results available early enough to influence the final bomber design. As a safety net for testing the wing, it was decided to build a second test plane as a backup so that the test program could continue uninterrupted in the event that the first plane was not available. Furthermore, the idea was to use the first test bed for evaluating the basic characteristics of the new wing and the second test bed building on the results to become the actual platform for probing high-speed transonic flight. Other than the negatively swept wing these two experimental airplanes, often referred to as "flying mockups" in Junkers documents, had nothing to do with the six-engine jet bomber later known as the Junkers Ju 287. A fact that has often been misrepresented in print. Since there was initially no official RLM designation available, and the later design of the fuselage - particularly in the cockpit area - leaned on the development of the Junkers Ju 288, the two test planes temporarily received the designations of Junkers Ju 288 V201 and Junkers Ju 288 V202. Indeed, there are many who even claim that the later Ju 287 had been developed from the Ju 288. This designation also served to maintain secrecy, for it gave the impression that it was actually a follow-on development of the Junkers Ju 288 and not in any way a new design altogether.

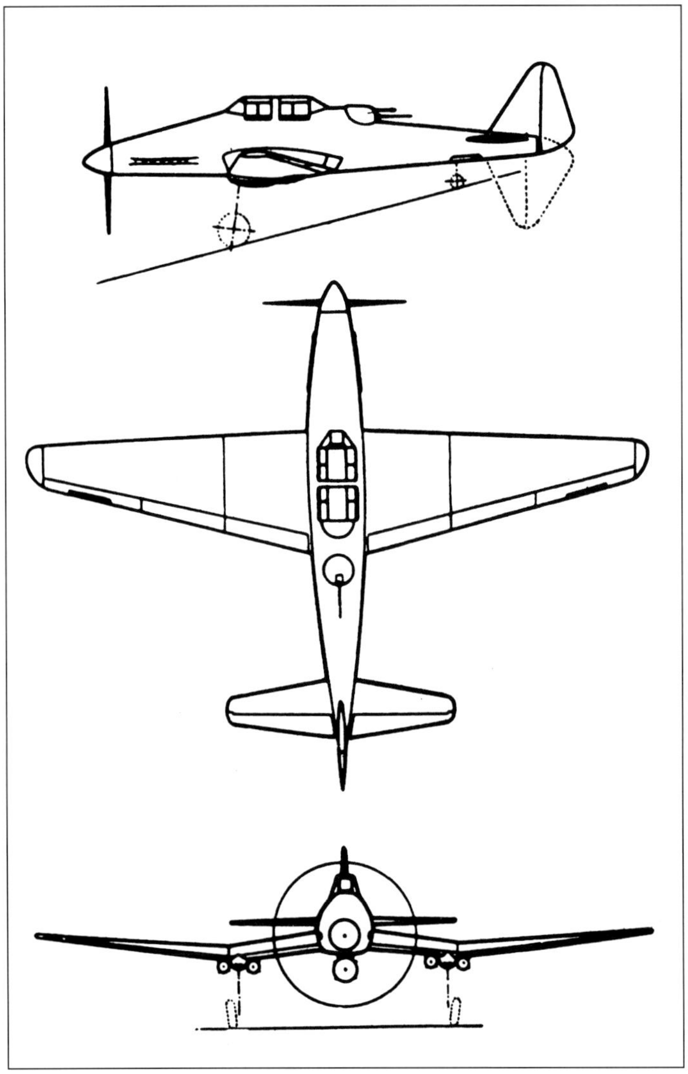

Three-view of the Ju 287 Stuka with the broken lines showing the rudder's downward fold, enabling a clear field of fire to the rear.

The later designation of Junkers Ju 287 would lead to further confusion since it was also an attempt to lead spies from the other side to believe that the type was a follow-on development of the Junkers Ju 87 Stuka. The flames were fanned by the fact that the Junkers Company had indeed been working on a more modern successor to the Ju 87 in the early 1940s that was initially given the designation Ju 187. The appellation was eventually dropped to be replaced by the Ju 287 design. Prof. Dr. Heinrich Hertel confirmed this information in a 1950 postwar interview. Development of this Ju 287 Stuka advanced to the mockup stage. It was a better-performing successor to the Ju 87 Stuka, incorporating aerodynamic refinements and other technological goodies. One curious feature was the option of rotating the rudder downward 180 degrees during flight in order to provide an optimal field of fire to the rear. In 1941 the RLM quietly canceled the program altogether following the poor showing of Stukas during the Battle of Britain, which in turn freed up project number 8–287 for allocation.

The following chapter takes a look at the developmental history in chronological order up to the war's end using authentic reports, with separate chapters examining the Ju 287 V1's testing, the V2 and V3 prototypes, as well as production and operational planning.

Above: Mockup of the Ju 287 Stuka. Of interest is the periscope mockup for remote control of the two MG 151s.

Left: Detail photo of the Ju 287 Stuka mockup's cockpit, showing the sighting mechanism for the gun position.

The fully-armed model of the Ju 287 Stuka.

Development of the Junkers Ju 287, Projects and Overview

Officially, development of the Junkers Ju 287 began sometime around the end of 1942 and initially manifested itself in the form of several developmental aircraft designs, or EF (for the German *Entwicklungsflugzeug*) as they were called at Junkers. Leading off the series of developmental design studies was the EF 55, an engine-less low wing monoplane with forward swept wings, followed by the EF 58, EF 59, EF 66, EF 67 (with two fuselage mounted engines), and the EF 68 through to the EF 116. The latter can be considered the great-grandfather of the later Ju 287. It was a four-engine design study for a jet powered bomber with negatively swept wings.

In order to better study the behavior of airflow along the wing an improved and much more detailed design was created under the designation of EF 122. Two wind tunnel models of the EF 122 were built in short order, each with a different engine configuration (see photos below and on page 23), and were used in quite a lot of tests. By and large, the results proved as good or better than expectations, with a slight advantage going to the engine configuration using two on the forward fuselage and two beneath the wings. This was undoubtedly one of the reasons why this engine layout found its way to the two later Ju 287 test beds. Other designs were drawn up and their relative merits/faults discussed. A few models were

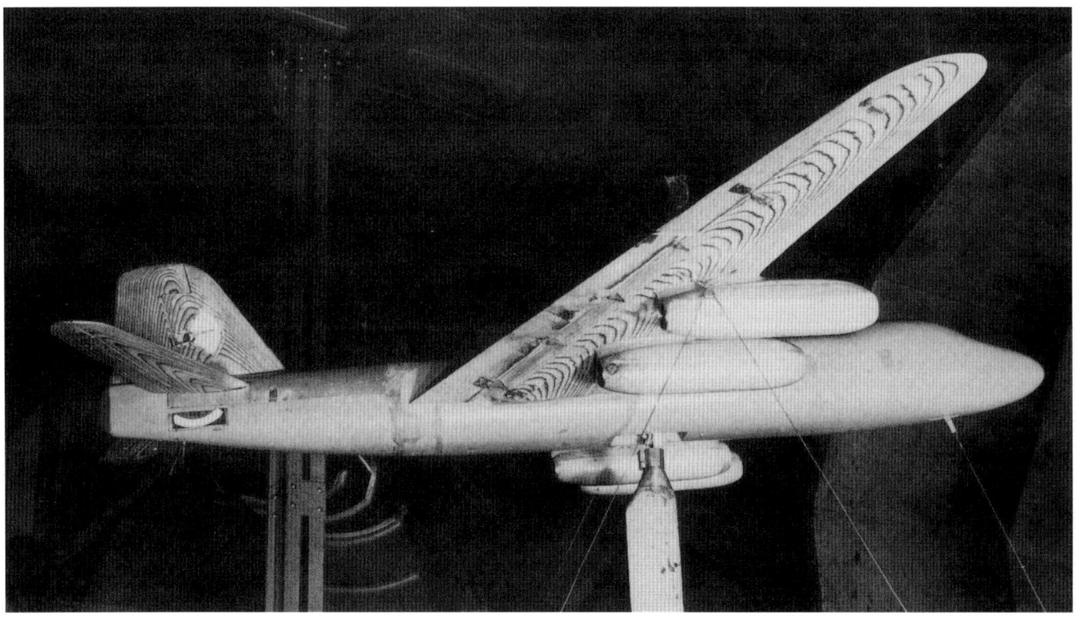

Wind tunnel model of the Junkers EF 122 with conventional mounting of the four jet engines beneath the wing leading edge.

Above: Another wind tunnel model of the EF 122 with the engine configuration the same as was later used on the testbeds, showing a fuselage mounted engine pod and a wing trailing edge mounted engine pod on either side.

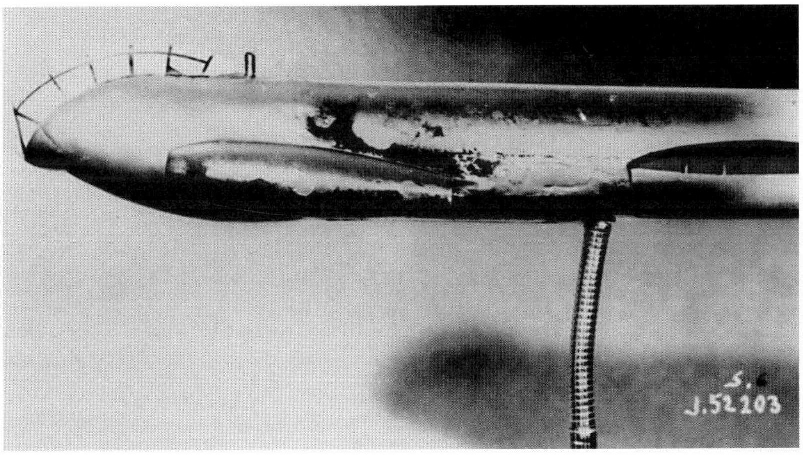

Left: Wind tunnel study with fuselage engine pods and the 360-degree canopy planned for the production versions.

The study's top view shows to good effect the lateral configuration of the fuselage mounted jet engines. Early wind tunnel studies already revealed the positive lifting qualities of this arrangement, which to a certain extent acted as canard wings.

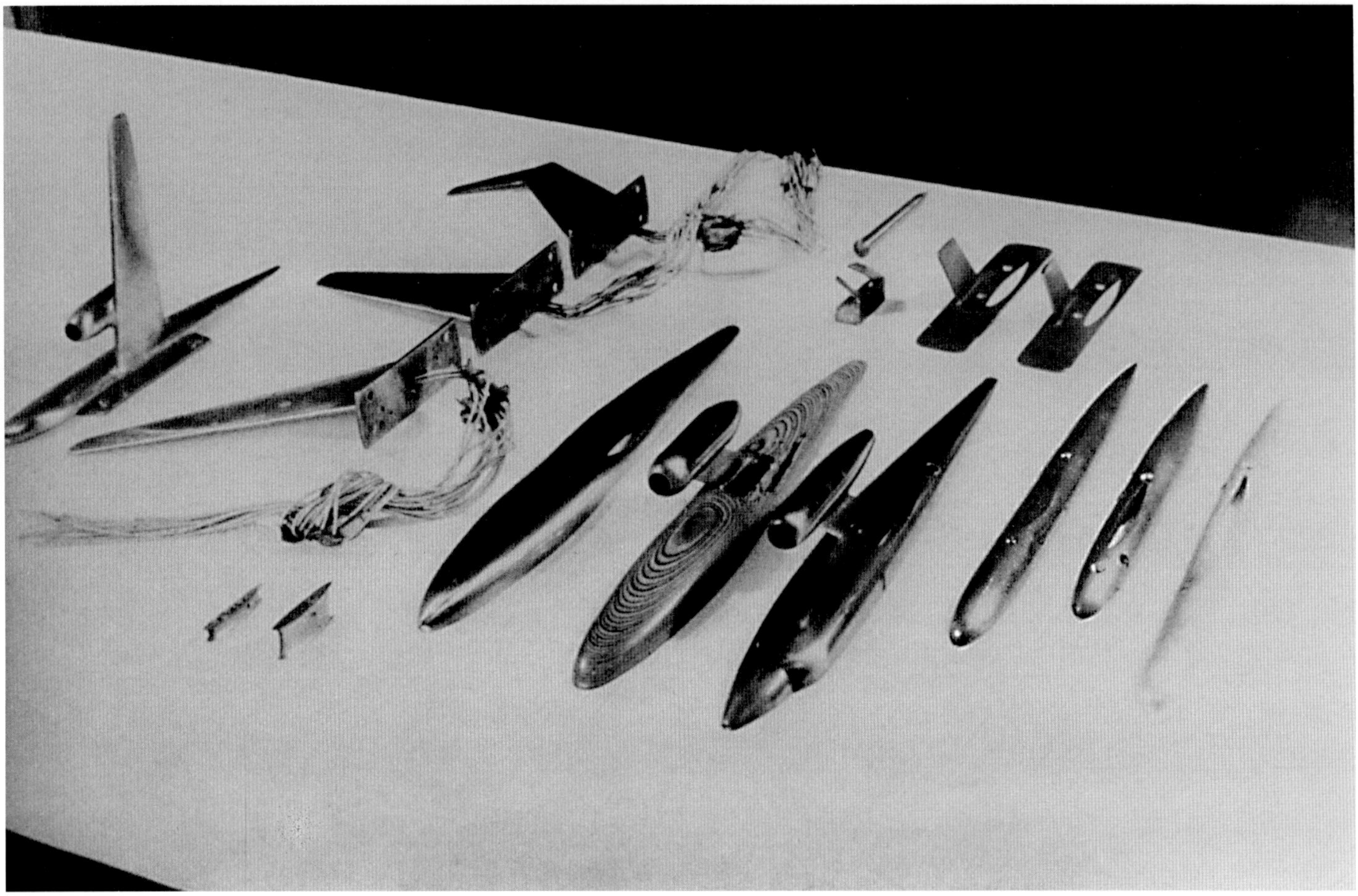

A colorful pallette of the wide variety of wind tunnel models, revealing the different wing designs, engine arrangements and fuselage shapes used in testing.

built from these designs and tested in the in-house wind tunnel in Dessau.

Some of the wind tunnel models were sent for testing to the *Deutsche Versuchsanstalt für Luftfahrt* in Berlin and the *Aerodynamische Versuchsanstalt* in Göttingen as well as the *Luftfahrtforschungsanstalt* at Braunschweig-Völkenrode. Ostensibly for follow-on wind tunnel testing, the real reason was most likely to have Junkers' own results officially confirmed. The wind tunnel trials and calculations continued until around the end of 1943.

A wonderful discovery in the field of aerodynamics resulted from the intense focus on the sweepback theory - the concept of so-called area rule. Area rule determines the most favorable location of engine nacelles in terms of airflow and drag. A design built with area rule in mind is shown in the drawing on the right. The discoveries born of these configuration criteria known as area rule, something that today is inherent in every jet powered aircraft, were patented on 21 March 1944 under the number 932410. Owners of the patent were listed as Junkers colleagues Prof. Hertel, Frenzel and Hempel. Along with the patent for "low-drag design of high speed aircraft, to include those with propulsion systems mounted outside the contours of the airframe", as the official patent text reads, was also the discovery of the so-called "wasp's waist" fuselage, also called the "Coke bottle" fuselage after the war. As can be seen, the idea was well ahead of its time and these discoveries were neither applied nor verified until years after the war. But now back to the year 1943.

According to conference meeting minutes from 18 November 1943, that autumn Adolf Hitler called for a push in the development of a jet bomber. At that time there were two developmental projects being bantered about, both with a low priority. The Heinkel project carried the designation 1068 (later to become the He 343) and relied heavily on the currently existing Arado Ar 234, a two- to four-engine reconnaissance aircraft and auxiliary bomber. Basically, the proven Arado design had been taken *in toto* and enlarged, resulting in a four-engine medium bomber in the weight class of 16 metric tons at takeoff. The Junkers team, under the direction of Dipl. Ing. Hans Wocke, developed an entirely new design and to a large extent in doing so had to wrestle with all the vagaries of un-

Model of the Ju 287 as idealized in the early stages of design.

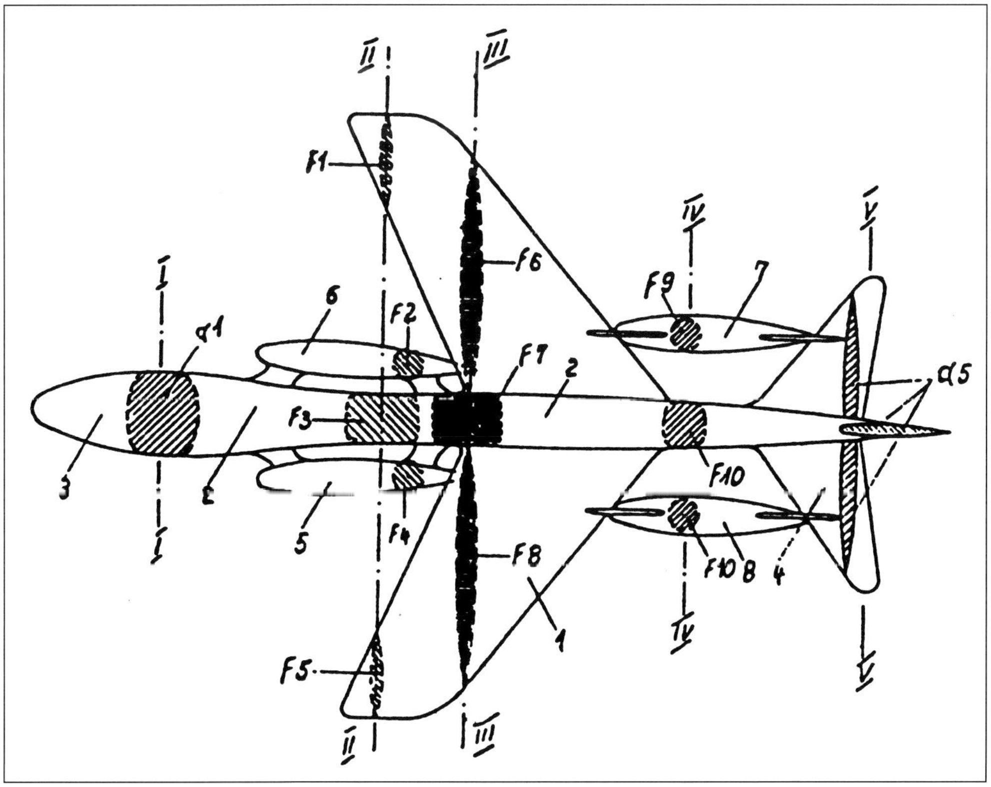

Excerpt from patent no. 932410 "low-drag design of high speed aircraft with propulsion systems mounted outside the contours of the airframe", from 21 March 1944. The sketch clearly shows the "Coke bottle" fuselage as well as the location of the engines on the fuselage and the wing trailing edges, ideal features of the area rule concept.

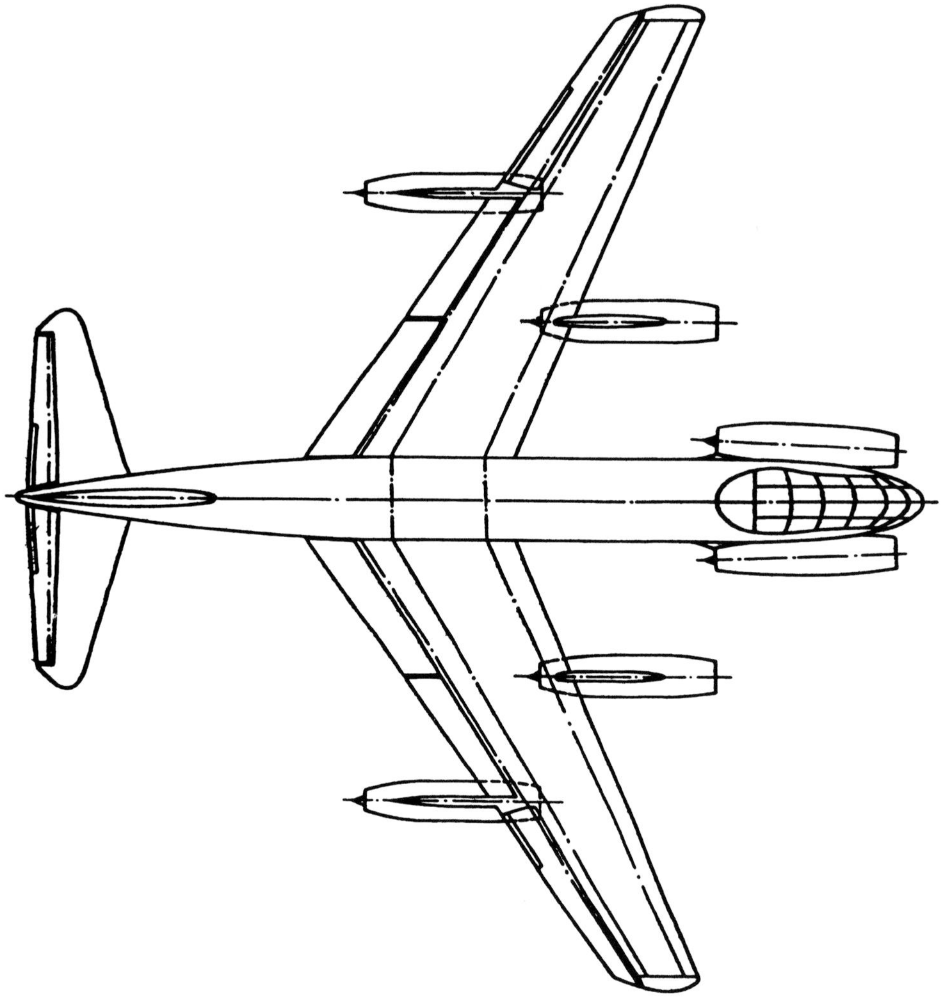

Rough design drawing for an ideal engine layout based on the area rule concept that, for technical reasons, was never realized (insufficient wing stiffness and poor power distribution through the wing).

charted territory. These included using six jet engines and pushing a machine weighing 22 metric tons to transonic speeds, just to name two.

The layout of the Ju 287 borrowed heavily from the Ju 288 already designed. The streamlined three-seat cockpit was used, along with the remote-controlled defensive armament as well as other design features of the Ju 288. No wonder that many have assumed the Ju 288 was the basis for the Ju 287's development.

During the spring of 1944 there were several fundamental discussions on the purpose, use and urgency behind a jet bomber, abbreviated to STRABO (from the German *STRAhlBOmber* - jet bomber). As mentioned earlier, there were two developmental designs running parallel to each other at the Junkers and Heinkel companies. This circumstance would lead to more heated controversy over the course of the developmental cycle, but this will be discussed at the appropriate point. A few of these highly interesting discussion minutes have survived and offer an incomparable view of the events that were occurring at the time. One such set of minutes are those from a meeting held on 18 February 1944:

Participants:
***Generalfeldmarschall* Erhard Milch** - *Generalluftzeugmeister* and Inspector General of the *Luftwaffe*
***Oberst* Siegfried Knemeyer** - member of the Technical Armaments Staff
***Oberst* Marienfeld** - Commanding General, Bomber Forces
***Ingenieur* Egon Scheibe** - Advisor for Bombers and Reconnaissance Aircraft with the RLM
Stöver - unknown function

Scheibe: About a year ago, when the turbojet engines were giving us the first tangible results, we began looking at planning a STRABO with turbojets. In doing so, we've gone down two paths. On the one hand, we've tried to find the quickest possible way, so that we've created a single-seat light bomber using the Arado 234. We've also explored the question of how a jet-powered plane can be used to cover as a medium bomber. (There follows the reading of the report.)

Gen. Feldm. Milch: Junkers still assumes that they'll go faster the higher the altitude, while Heinkel will stay the same or even possibly go slower.

Scheibe: That is partly a matter of calculations, and partly a result of the design. We've had some wind tunnel testing done. This data was taken based on these tests, and we don't yet know how they'll translate into reality. Basically, though, we can say that the Ju 287 is expected to have better speed.
(**Gen. Feldm. Milch**: What is the service ceiling with tanks half empty and after dropping its bomb load?)

For the 1068, this is 11.5 km (author's note: by this he means the later He 343, which was initially carried under the project number of 1068 and for awhile was turned over to the DFS for design work) and 10 km for the 287, with the former based on the Jumo and the latter on the BMW engine.
(**Gen. Feldm. Milch**: What is the speed at altitude?)

805 km/h at full throttle at 10 km, at 6 km altitude it is 860 km/h. Maximum speed is at this altitude, then it drops off.

Knemeyer: The differences in speed aren't as great as with the O-engine (meaning an Otto, or piston engine).

Gen. Feldm. Milch: 50 kg bombs aren't a practical bomb load for the Heinkel, are they?

Scheibe: That depends on whether the standard racks are used or whether we intend extra equipment. That hasn't been worked out yet.

With regard to Heinkel, it needs mentioning that a version is being researched for the second stage which would offer an increase in speed at higher altitudes. This would be accomplished using swept wings. Model planes are being built that will be used for several tests. In addition, we can use the data on wing sweepback that we've already got from Arado.

Knemeyer: I've just got a couple questions to bring up about the projects that will need simultaneous decisions made. Based on what Scheibe has presented, these two projects have been laid out based on certain points of view in order to be optimally tailored to these classes. A decision needs to be reached on whether both projects can run at the same time. If this can't happen, then we need to decide which of the projects we should pursue for full-scale production and what we're going to do with the other one. Before we deal with these matters, I would suggest that the General of the Bomber forces tells us which appears the best STRABO to him.

Marienfeld: What's the wheel track of the 287?

Scheibe: It's much bigger than the Heinkel. The Heinkel has 2.80 meters, while the 287 amounts to 5 meters.

Marienfeld: I've only been able to take a quick look at the two projects, but I can see that the Heinkel project basically offers no advantages over the 234 if you compare all the data against each other. On the other hand, the 287 offers improvement in view of payload capability. I have here a 3 t payload potential with a number of load-out variations, while all the other machines I've had up to this point, to include the Ar 234 as well as the Heinkel, only have a 1.5 t payload capacity. With the Heinkel I can carry two bombs, but even then it's not clear whether I will be able to carry two 1000-ers. Probably four 500-ers will go inside. In a standard scenario, the two 1000-ers would be the better choice.

Then there's the comparison between the Heinkel and the 287 with regard to takeoff and landing characteristics. Here the 287 clearly suffers from a disadvantage in that it needs a large takeoff run. It needs 1,800 m compared to the 1000 m for the Heinkel machine. It's still up in the air how much this data will hold up in practice, though. Using RATO packs the Ju 287's run can be reduced to 1,100 m. But this brings up the question as to what type of assisted takeoff would have to be used, whether a new kind or the current kind, and whether it would be possible to attach them to the wings.

The landing stretch for the Heinkel is about 1,400 m, about the same as for the Junkers. In any event, for both machines the rollout length can only be reduced using braking parachutes.

The matter of tail armament will probably have to be resolved in the same manner for both machines. One option would probably be a fixed gun mounted in the wing trailing edge.

However, the 287 offers a significant advantage over the Heinkel machine. To be precise, I can fit a three-man crew in it.

(**Stöver**: Both compartments have been configured for a three-man crew!)

The Heinkel machine is 1.80 m at its widest point. How wide is the 287?

Scheibe: It has about the same dimensions. It is 2.05. This fact is determined by the requirement for the ability to carry the FX internally (author's note: FX is the acronym for the SC 1400 "Fritz" remotely-guided bomb.)

Marienfeld: The key factor playing a role here is the ability to carry a large bomb payload with several different load-out options. In this, the Ju 287 offers a major advantage over the Heinkel machine. I don't see any big advantage in the Heinkel plane over the Ar 234.

(**Stöver**: It has double the payload with the same range and an extra crewman!)

Knemeyer: I would now like to ask the General of the Bomber Forces to state what he feels to be more important: the larger payload capacity with the same range, or the extra crewman? The General of the Bomber Forces considers the Ar 234 to be available and says that he would like even more payload.

Marienfeld: I have both double the payload and the Ar 234, with peoples' minds made up on how it is going to be used. This means that I have another variant for operations.

(**Stöver**: And at the same time the all-weather operations are better covered.)

Gen. Feldm. Milch: The main question for us, Marienfeld, is that you tell us which variation you support.

Marienfeld: I would like as big a payload as possible with the same range. And the 287 would guarantee me that.

To be absolutely sure about this we'd need to have an accurate table that would show a comparison of all the machines we're looking at - He 1068 (the later He 343), Ju 287 and Ar 234, listing payload, crew, speed, range etc. so that we could weigh them against each other.

It just appears to me that the Heinkel offers no improvement on the Ar 234, whereas I see a big advantage in the Ju 287.

Gen. Feldm. Milch: I propose that we give Marienfeld a little bit more time so that he can inspect the two items at the companies.

Marienfeld: It is also important to see in practice how the bombing would go. Can the bombardier effectively carry this out or not, or would the pilot have to do it? In this case, I would need to have an automatic pilot.

Stöver: Both types have it, but on the Ar 234 the pilot needs to function as the bombardier as well.

Marienfeld: Then the cockpit needs to be big enough so that when I look at all the potential radio equipment I've got to put in, that the three crewmen who will then be required can have enough room to move.
(**Scheibe**: The compartments are exceptionally roomy!)

Knemeyer: Certainly, and in point of fact should not be compared in any way with the cockpit of a piston airplane since the large fuel quantity can be carried in a much larger space.

Marienfeld: Neither project has any armament. That also needs to be taken into account. The most important thing is to figure out where the guns go.

Scheibe: In both cases it is possible to fit a tail turret with modifications to the fuselage, and this turret can be controlled remotely by the pilot via periscope. But we don't want to bring up armament at this early stage of our thinking, for the simple reason that in the beginning armament won't even be necessary because of the high speed involved. We've only got to be ready to quickly incorporate it in case it proves necessary.

Marienfeld: For the Heinkel machine, the total weight of equipment, armament, etc. is expected to be 1,080 kg. I think that this weight is too low, and that we won't be able to manage this.

Scheibe: We've got to take a look at and work out the details. But we've got to apply a very rigid set of guidelines, because these planes are extremely sensitive to weight due to their long ranges.

Marienfeld: Then I've got to take a look at the details and figure out how much of an advantage the other project is to me. If, for example, I've got to do without two guns to keep the weight at 1,080 kg, then I can't use the machine; if I have to do without the third crewman, then the disparity between this machine and the Ju 287 becomes even greater.

Stöver: You can't have a direct comparison of the two machines, since the Heinkel is a quick-fix solution with minimal risk involved, while the Ju 287 won't be appearing for another one to two years.

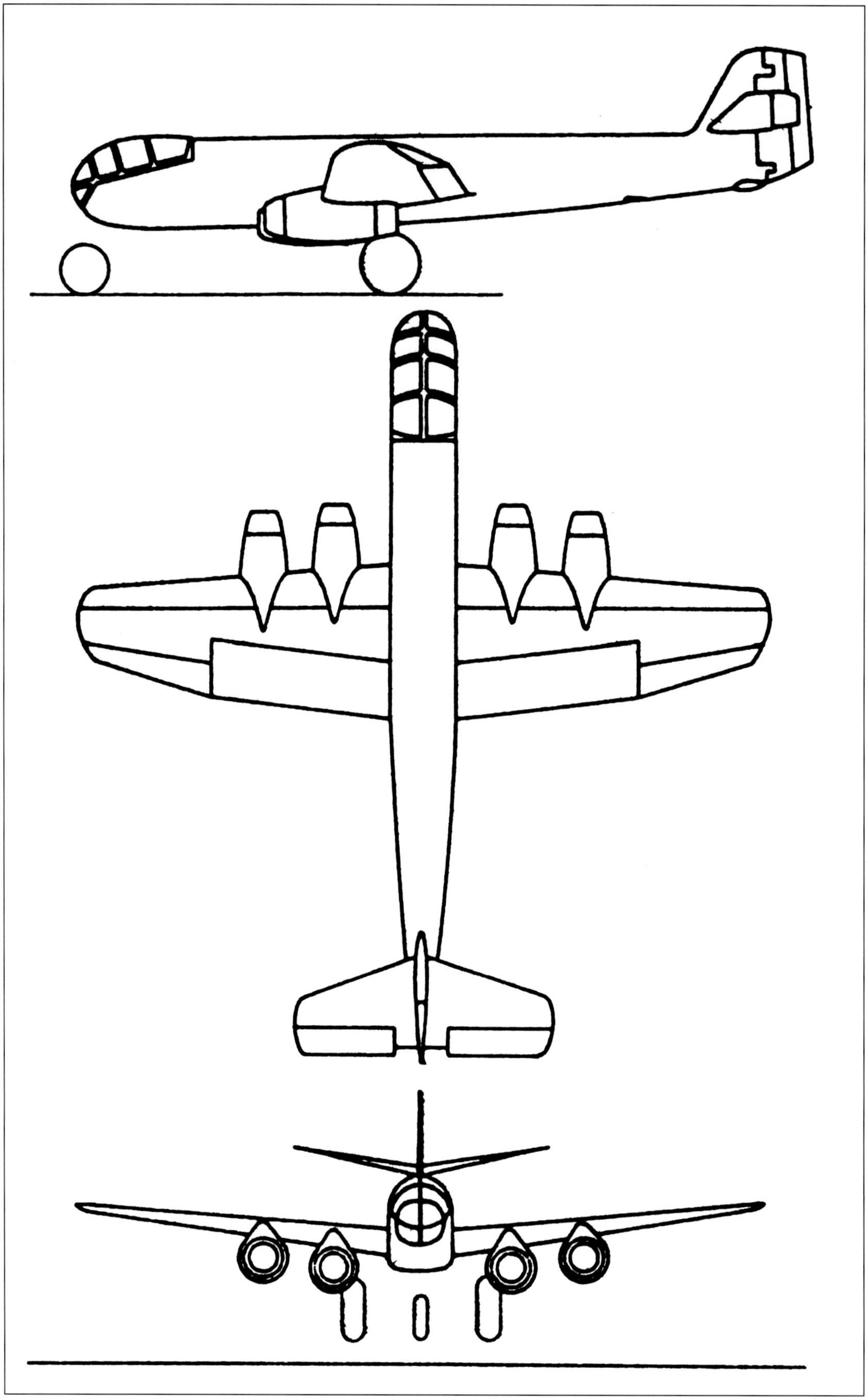

Three-view of Heinkel's highly refined competing design, project 1068, which later evolved into the He 343.

Gen. Feldm. Milch: We can't push the issue here today as it's not definite and we shouldn't be discussing it. The time you'll need will of course be given to you. I believe the question is: can we carry out both projects simultaneously? I believe that this is not possible. For the time being, the Heinkel machine will be covered by the Ar 234; on the other hand, in the war against Britain we aren't covered in the area that we would be with the Junkers machine. The Ju 287 is actually a typical bomber for relatively short ranges. It has a respectable speed, a noteworthy bomb payload, etc. I see the Heinkel machine to be first and foremost a good *Zerstörer*, something that we're lacking in the foreseeable future, and one that can carry a limited number of bombs. But this doesn't meet the tasking for retaliation against England by bombing London. Instead, the machine would play a role were any landings to take place somewhere. I feel the Heinkel machine to be the more flexible type that's worth considering for *Zerstörer* missions. In the tactical arena, I see in it the replacement for the Do 335. The Ju 287 is a bomber, period. The He 1068 is suited primarily as a *Zerstörer*, assuming it can carry out *Zerstörer* roles with a bomb payload. But for it this bombing aspect is of secondary importance for now. The Heinkel project has to be worked more with clear-cut objectives set, with the General of the Fighter Forces having a say as well. The Heinkel machine has the advantage of having lower weight and so is in the foreground when it comes to climb rate, takeoff run and service ceiling. This is something that can't be overlooked for many roles. Data for the Ju 287 is here only to support specific large scale bombing operations, but not for general combat operations.

The Ju 287 is unsuited for *Zerstörer* missions. The Heinkel machine can actually offer us more in this area.

Scheibe: I'd like to add that Heinkel is researching how to fit the Ju 287 with armament.

The four-engined version of the Arado 234 - here an Ar 234 C3 - served as a stopgap measure for the anticipated Ju 287.

Gen. Feldm. Milch: The Generals of the Fighter and Bomber Forces will have to deal more directly with both projects within the next few days in order to give their opinions. In particular, they need to focus on bomb load, fuel tank location, etc.

I suggest that the General of the Bomber Forces state when he thinks he can be finished with his verdict on both projects. Work should continue at a steady pace on both projects for the time being. All remaining concerns basically have to do with establishing a central focus. Especially important is, of course, the question of range, the question of speed and that of armament. These are all matters for the generals to consider.

Marienfeld earlier proposed the idea of a comparison chart for the aircraft types under review. It would be good to also include data for the Me 262.

Other problems with the Do 335, Fw 190 and He 219 were discussed over the course of the meeting, which ended at 15:15 hrs without any real decisions being made. The meeting minutes clearly show the recognizable disagreements between the advocates of the Heinkel and Junkers design at this early stage of development.

Seven days later, on 25 February 1944, *Generalfeldmarschall* Milch called another meeting with a slightly different lineup of participants. Since this meeting played a decisive part in the further development of the Junkers Ju 287, it has been reproduced word-for-word below.

Participants:
***Generalfeldmarschall* Erhard Milch** - *Generalluftzeugmeister* and Inspector General of the *Luftwaffe*
***Oberst* Siegfried Knemeyer** - member of the Technical Armaments Staff
***Oberst* Marienfeld** - Commanding General, Bomber Forces
***Oberst* Edgar Petersen** - Chief of all *Luftwaffe* Test Centers
***Dipl.-Ing.* Karl Frydag** - Director of aircraft production programs on the Armaments Staff, General Director of the Ernst Heinkel AG and Technical Director of the Henschel Works

Gen. Feldm. Milch: Then to the next point, *Oberst* Marienfeld. Decision on the STRABO development.

Marienfeld: If we compare the two projects, the Ju 287 and the Heinkel, we can only do it by taking into account their origins. On the one hand I had two different projects, taking the wing from here and the fuselage from there and thus creating a combination, while on the other hand the Ju 287 was a design whose maker had carefully thought-out its development and was complete in every detail. I'm not saying that this was the reason that the combined design was lacking in many areas (author's note: the terms "combination" and "combined design" refer to using the Arado Ar 234 as the basis for the Heinkel design). But when I compare the specific data side-by-side, there are quite obviously some very noticeable differences. What I must insist on is as large a bomb load capacity as possible on the one hand. I really want a plane that can do the job. I don't agree with the viewpoint that I can use the STRABO in poor weather and at night. It depends on a payload as large as possible as well as much potential variation in the payload capability in case of concern with range, plus the option for armament.

If I look at bomb capacity, with the Junkers project I get a load of 3 tons, or 2 tons with the Heinkel project. For both of these projects it's possible to increase the load by one ton each by dropping a corresponding amount of fuel - for Junkers this additional ton could be carried somewhere in its well designed bomb bay, for Heinkel it could only be by adding the one ton externally. Since it couldn't be added underneath the wings because of the number of jet engines there, the load would have to be carried under the engine pods looking like a cluster of grapes. In doing so, I've found that the speed would be reduced by such-and-such an amount and that its climb performance would also drop correspondingly. Looking at the overall picture, I can't really exercise these several options when it comes to bomb load.

With regard to range, it can be said that I've found Junkers to give a total of 460 km/h more; with later upgrades, i.e. next spring when we have the Heinkel engine, the range would increase even more by 600 km/h. With regard to speed, the data are the same.

Basically, it can be said of the Junkers project that I've got a completely satisfactory range even when the payload is increased from 3 to 4 tons, while with the Heinkel I've got a limited range and reduced speed. If we compare the range data against each other, there is only one way you can say that if I've got to trade off fuel for bomb load, meaning I'm forced to carry less, and that is when level flight - when we first see the high speeds - is compared. And it's precisely this damaging factor - a range reduction in level flight, i.e. in other words, the ratio derived form the overall range is in real-world operations not as good as for an airplane with a lesser range that I actually have to deal with.

Regarding armament: the Heinkel might possibly be fitted with fixed rear-firing guns, the Junkers even has the potential for a type of gun turret installation. For both of the projects, the takeoff run is only the same if I use assisted takeoff for the Junkers. The fitting can be done without problem. Both data for the landing run are virtually identical. If there are certain differences in the landing run it's because of the use of a blanking 'chute (author's note: this refers to a smaller version of a parachute brake known as a ribbon braking parachute), and I would like to point out that we don't yet have any experience with the smaller 'chute.

With regard to the undercarriage I arrive at 16 to 18 tons for the projects. For the Heinkel I have a wheel track of 2.16 m, for Junkers it's 5.60 m. Up to now we don't have any experience as to whether a fully loaded plane landing on rough terrain would be in danger of tipping over. During takeoff there would probably be no problems on the runway, so that we could possibly base our conclusions regarding this point about the landing gear on the Arado 234, although on a smaller scale. In reality, I'd like to say that I have certain reservations about using this narrow track undercarriage with such a large overall weight.

All in all, I can say that the Ju 287 offers me significantly more advantages as a work plane based on the specific data I've just presented, so that I personally have decided in favor of this project. Both would probably go into service at about the same point in time. For the Heinkel, if the necessary materials were made available production could probably begin in the spring. For the Junkers, this matter hasn't been discussed in detail, but in the view of Professor Hertel it would be the same, after he's given us the individual deadlines for continuation of development - i.e. sometime next spring as well.

Knemeyer: The origins of these two jet bombers, which are a step beyond the Arado 234, are such that the various points mentioned by the General of the Bomber Forces as being to Heinkel's detriment when compared to the Junkers only came about because of their origins. Earlier Heinkel was also in the 3 ton class, the same as Junkers. Through an agreement between the two companies it was decided that only Junkers would continue working on a design in this size category. As a result, Heinkel continued to work on its project, but redrew the design for a 16 ton class with a view toward developing the design in such a way that there would be no risk involved with regard to flight handling and such things. At the time we were of the opinion that it would be possible to start work - at least developmental work - on two jet bombers at the same time. It was divided like this because it had since been determined that starting two of the same class is not a possibility. In the meantime, we've discovered that a decision must be made whether or not we can focus on both of them at the same time, and we'd planned on the General of the Bomber Forces deciding on which of the two projects - if it comes down to just one - he would want to have. I think that we don't need to bring out so much the pluses and minuses of this project or that one, because these are determined by the history of their origins, rather we need to emphasize the requirements outlined by the General of the Bomber Forces. In any event, he wants a bigger payload. His biggest priority, if we go beyond the Ar 234, is a bomber that can really carry a load. As it's been stressed here, the General of the Bomber Forces says that what's won him over is this massively large breakout figure of 3 tons, even 4 tons in several load options. I would personally say that the other design of Heinkel's must continue developmentally with the same emphasis, but fundamentally as a potential heavy fighter, night fighter and reconnaissance and such things. Because it's in this class of 15 to 16 tons that we'll later find the proper replacement for the Arado 234, which in any event can only be considered a stop-gap jet bomber.

Petersen: I'm of the same opinion, that we've got to gamble on the 22 ton bomber Junkers project and go with the larger payload from the outset.

Gen. Feldm. Milch: Only that development should continue on the Heinkel project.

Knemeyer: I would recommend that it continues with the same emphasis, only not be rushed into immediate production, just developmentally with the same emphasis. It's been shown that we'll get the real work bomber in this configuration that will result in a standardized fuselage that would cover the heavy fighters and reconnaissance planes and everything. I think that ultimately, if Heinkel is left alone developmentally, we'll get more with regard to speed and range.

Gen. Feldm. Milch: This means that it should fill the gap that is now being temporarily covered by the Arado 234.

Knemeyer: Certainly. It would mean continuing with both projects with the same emphasis, but production would first begin with the Junkers and then later for the Heinkel project.

Gen. Feldm. Milch: There's another view that the General of the Bomber Forces brought up yesterday, and that is that the only four-engined bomber we've got is the Heinkel 177 and that there's still a lot of work that needs to be done with it. The General of the Bomber Forces has called for the same priority for this. And if the General of the Bomber Forces has decided on the Junkers jet bomber, Heinkel would place even more emphasis on the 177.

Petersen: I've flown the four-engined 177. I am astounded at the flight performance. It is a machine in the same class as the Junkers 188. With boost pressure, the machine does 440 km/h at sea level, has a great climb rate and a good takeoff and everything. The second point is that IX Corps swears so much by the 177 for its England operations that General Koller says the command staff is of the opinion that the 177 is the ideal machine for London operations, since they've not had a single loss over London. You can say that the 177 B 5 is a machine that tactically and technologically we've gotten right.

Gen Feldm. Milch: Perhaps Frydag offer his opinion on the matter from a Heinkel and Industrial Council perspective.

Frydag: I can't do so well from a Heinkel perspective, since someone from Junkers should also be sitting here. To that, I'm of the same view as Knemeyer. There are to widely different roles. The Junkers machine is a heavy machine, the other a medium one. If we could afford everything, we'd have to say: we need Arado 234s, the Heinkels for the medium class and the Junkers for the heavy class. If now it's a matter of heavy or light, then these are purely tactical considerations and you can't even begin to address these from a technical standpoint. That's the question: does the front need the heavy one or the light one. However, if the front comes down for the heavy one today, then I would issue a warning. The heavy machine isn't here yet and I was, speaking for Heinkel, quite happy when Heinkel was freed from building the heavy machine and could work on the 16 ton machine. Because the way I see it, the 22 ton plane is not here yet. It would be entirely wrong to make the following decision: bet everything on the 22 ton machine and leave the 16 ton plane, whether Heinkel's or Focke-Wulf's, in the background. No, you've got to really push up the tempo on the medium's development in order to have any type of result as quickly as possible, and at some point you've also got to be clear on whether or not you want to go into full production. We can't wait until we've got testing results.

Marienfeld: I agree with *Herr* Frydag, the development of the Heinkel should not be stopped. It may be that you'll get good results with the new Heinkel that Knemeyer likes, and that you maybe can use this new airplane as a basis for other, more powerful and heavier bombers. Today it depends on me getting a work plane in the shortest time so that I can gain the upper hand over the enemy. I've got to take advantage of this.

Knemeyer: I am certain that we urgently need the medium design. But the General of the Bomber Forces says that he's already got a jet bomber with the Arado 234, beyond that he wants more load capability. It's clear from a developmental standpoint that the medium design is well suited as a standardized airframe for all weapons platforms, and therefore it needs to continue developmentally with the same urgency. Perhaps it is good if Heinkel only works on the development for now and focuses its labor behind the four-engined 177 and we wait and see what kind of flight results the Arado 234 gives us.

Gen. Feldm. Milch: In conclusion I would like to say: the underlying result is actually something I didn't want to bring up in our earlier meeting. I see the matter as this - it is not a purely tactical question. The technological side also needs to be turned on. I can't say anything different than what the General of the Bomber Forces has said, precisely because the stop-gap solution from Arado offers nearly the same low-level performance with the same range at about the same speed. Compared with the Arado project, the Heinkel project would only give us a specialization in the bomber category as it now stands and as it would have to be built. Therefore, I believe that tactically you have to say that for the General of the Bomber Forces the first area isn't so good with the Heinkel proposal, that the second area must be covered, a real heavyweight, a bomber, with the matter of range being the most crucial for me. Since the combination between low-level and high-level flight makes a big difference in range, at the least at a ration of 1 to 2, I think that the role the General of the Bomber Forces wants can only be covered with regard to range by the Junkers proposal. We can't just have our eyes set on the London target when we're going long range, we've got to look at targets even farther away as well. We've got to apply it to other theaters as well, we have to think of Italy, of the Balkans, of the East. I believe that for tactical reasons there can be no other decision.

Technologically, I think that the risks inherent in both proposals and that have been presented to me in different ways are not as different as they've been made out to be. In both cases it comes down to the engine type. I don't see any major problems in the proposals with regard to airframe and equipment. I know that the gentlemen here may disagree with me somewhat on this. I see that the swept wings, etc. offer an opportunity for study, an empirical potential solution after the wind tunnel tests that will nevertheless culminate in a practical solution. Whether the sweep is forward or rear, greater or lesser, are matters that we can and must take the opportunity to resolve along the way, since we will need these answers if we are to concern ourselves with Mach numbers. I can't say whether that will hold things up longer. I don't think that it will hold things up by much, because preliminary testing should be done using piston-engined planes as well, and this can begin right away. I'm convinced that wing sweep, when taken overall, will somehow offer us a critical improvement in certain areas. The Lippisch side is making great effort to clear up the matter of wing sweep. Therefore, I see the overall technical risk actually - and I'm using round numbers here - at 70% to 80% on the part of the engines.

Frydag: That's what I wanted to say. This makes the Junkers proposal the bigger risk, as long as it has the same power engines.

Gen. Feldm. Milch: Which is why the Junkers proposal absolutely needs RATO, etc. The General of the Bomber Forces is fully aware of this.

The third point is: I ask that this is made clear from our earlier meeting. That way the Heinkel proposal doesn't get swept under the rug, but instead get time to really build on the experiences of Messerschmitt, Arado, and also Junkers as much as possible in order to then provide a superior solution in the medium size category., because the medium solution can provide us with a joint solution for bombing tactical targets as well as for heavy fighter roles. The Heinkel proposal, then, needs to stand out from the other one because of this dual application on the one hand, and on the other hand it needs to give us more of speed, climb rate, takeoff speed, service ceiling than the other one. If I look at the other side of the coin, that one day there will be jet fighters, etc., then we'll see the liveliest race for speed break out; because we'll once again have the situation where the fighter can essentially fly as fast as the bomber, but in fact will have to be superior. We've always operated on the assumption that the fighter is superior to the bomber. For the Heinkel project we've therefore got to look at the matter of speed as a priority at the expense of bomb carrying ability, at the expense of range, a very difficult matter since this clearly puts us in the realm of Mach effect. Therefore, it would be a good thing if we had some time for this solution, to advance in this area, and even use other experience in conjunction with this. The fighter will also initially find its limit somewhere with Mach numbers. The question is, whether or not we'll succeed in finding a constructive solution that can make some type of headway in this area, hopefully more than the enemy can in his first attempts. Therefore, the issue for the Heinkel medium bomber proposal is entirely positive. It's got a long way to go yet, if it's laid out the way I've said, it's got more prospects than the fighter project. Technologically, it's got more problems because of the issue with supersonic speed. We can't rush into this blindly, but we've got to limit ourselves and say: that is the most we can give. In addition, we've said today with full justification that the four-engined Heinkel 177 solution has to continue as a key focus of our task, because this machine will continue to cover our needs. Part of this issue is that we won't be jumping into jet power on a wide basis, which will make a sensible replacement of the old piston-engined aircraft types easier. However, once we're safely up and running with jet powered flight we won't have a problem finding the capacity for manufacturing, because the entire Junkers 88, 188 group that will continue will need a replacement and this replacement, as Knemeyer points out, is best found in the 16 ton group. 16 tons have always played an important role for us. The requirement has always proven a good one and the right one and whatever happens we'll have the capacity in this area, so that maybe we can later see the final results; the Junkers proposal will go where the 177 is being built, as it were, and we can use the Heinkel proposal where the 88/188 is being built.
(**Frydag**: That would be grotesque!)

That it would be, but it would be the right way. Which companies build what has not yet been determined.

Frydag: For the continuing future of the medium approach, I see the following big advantage. The medium solution that's now being offered with Junkers jet engines will have get an enormous boost the moment the Hs comes available (author's note: he is referring to the He S 011 Heinkel turbine), whereas the Junkers proposal will only get this advantage when the next stage larger BMW is here. That's a long way off still. To this extent I see here a great development potential for the 16 ton machine. The right way would not be to approach it the way you've outlined it.

Marienfeld: The technical side was the deciding part for me, the work side. I have fundamental issues with a 16 ton design having a wheel track of 2.60 meters. You've only got to take a look at the Me 109 compared to the Fw 190. For this reason alone the wheel track would need to be much wider. If I have 16 tons, on landing 8 or 9 tons, then I've got to express my reservations about this little wheel track. This will crop up with the general of flight training as well as on the front.

Frydag: That is certainly true. But there's a new problem that's come up with the landing gear: there's no longer an engine nacelle to retract the undercarriage into. Here we have something hanging below that is filled with a jet engine, so that I've got to hang the landing gear out on the wing someplace. Whether or not you'll manage to get a wing with a retractable undercarriage is a problem. Both Heinkel and Arado made the narrow track gear in order to get it into the fuselage.

Marienfeld: If I have an undercarriage width of 2.60 meters, than I'll probably be forced to make use of powerful brakes if I want to taxi on the ground. I don't know if there's additional data in some other areas that can't later be resolved from a technological standpoint.
(**Frydag**: There's no history data available)
When the matter of the 287 is resolved, I'll take this one.

Gen. Feldm. Milch: The best opportunity to resolve all these wishes and questions is to tackle the Heinkel project with energy while we've got the time. We have time to really bring something to fruition here.
(**Marienfeld**: We've thoroughly discussed the project from every angle and are convinced that the 287 will be here within a year)
If anyone has any different opinions here, then please speak up. - This is not the case.

With this the STRABO theme was abandoned and the discussion turned to other problems. The meeting minutes show that at the time there was a real problem articulating understandably. Also, we can see from the discussion that there was most likely already an awareness about the physical and technological problems with transonic flight. On the other hand, it is amazing that they assumed there was plenty of time remaining for developmental/technological problems to be resolved for the Heinkel proposal, a view that was clearly out of touch with reality given the stage of the war at the time. Furthermore, it is interesting to note the value placed on the jet engine's development, which at this point was still quite unsatisfactory. Another item worth noting is the fact that both the Heinkel company as well as the Junkers company were independently developing their own jet engines and that a hotly contested competition developed between the two in this area. Yet from the outset the Heinkel company understood that the idea was to have the Junkers design ready for frontline operations as soon as possible, whereas the Heinkel company was only expected to patiently work in peace toward developing an optimal design.
Accordingly, *Generalfeldmarschall* Milch stated during the course of the fighter staff meeting on 26 May 1944:

Yesterday we discussed the matter of the 287, the jet bomber, with the *Reichsmarschall*. By June/July there will be a pre-production series of 100 machines. By December monthly production will start running at 100 units, and this we're clear on: the developmental airframes can partially, but not all, be covered by the aluminum stockpile reserved for development. The bulk must be covered by the types we're dropping, meaning from the bomber sector. This can only affect the Ju 388/Ju 88. It is a question of deadlines here, when we can have the Junkers, and just how far along the machine is ready for production. The views Hertel expressed yesterday were quite positive. Hertel, however, hasn't had much luck with any of his latest designs. Naturally, I don't doubt his abilities. But his poor luck comes from the fact that his designs were dropped when faced with the competition. Therefore, we have to pay attention and not blindly trust in this and say: it will be here!

I only wanted to mention this so that it can be factored in. Work on the 287 and the 177 has got to be stashed away, this is the important thing.

This excerpt shows that the push was for the Ju 287 to go into production as soon as possible; on the following day there inevitably followed the finalized construction contract issued to the Junkers company for an initial batch of 21 aircraft for flight testing and as pre-production machines. This included the first three aircraft as pure test beds, with the remaining 18 to serve as pre-production prototypes.

Following a series of project studies the decision was made to lay out the finalized version of the Ju 287. Accordingly, the Ju 287 V3 would be the pre-prototype and the Ju 287 V4 would serve as the prototype for the expected series production. Both the V3 and the V4 were to be powered by two triple engine combinations using BMW 003A engines, each with 800 kp static thrust, giving a total thrust of 4800 kp.

After the first detail designs for series production began, in the background there continued to brew the competitive conflict between Heinkel and Junkers. Heinkel felt as though he'd been treated like a stepchild and felt it necessary to compose his displeasure in a memorandum that he drew up and published on 16 June 1944:

Author: Prof. Dr. Ing. e. H., Dr. phil. h. c. Ernst Heinkel
Memorandum on the He 343 and Ju 287

I. Tasking

When in January of this year we were tasked to design a bomber borrowing heavily from the Ar 234 jet fighter, build 20 prototypes and at the same time prepare for its full-scale production, the following considerations were levied on this tasking

1. It is necessary that the *Luftwaffe* have a frontline-ready jet bomber available as soon as possible.
2. In order to obviate any risk involved with a rapid startup in production, the design must be based on proven principles.
3. Choice of engine must also be made from this viewpoint. Therefore, only those jet engines actually available and already proven are to be fitted. Turbojet engines with higher performance but that are still in development or that have not fully proven themselves can be considered for further development of the type.
4. Parallel to this tasking, Junkers was given the contract for design, prototype construction, and large-scale production of a jet bomber, in which a series of risks would consciously be accepted. In the best case scenario, the risks associated with a forward swept wing offer a speed improvement calculated at about 30 km/h, something that for our part does not appear to be worth it for a type expected to be built in short order.

II. He 343

1. Design
   a) In the area of jet bombers, we have already carried out valuable preliminary work with our efforts on the P 1068.
   b) In order to obviate any risk involved with a rapid startup in production, the design was based on a 1.5x enlargement of the Ar 234.
   c) 4 Jumo 004 C engines were selected as the powerplants, each with a static thrust of 1,000 kg. This engine has already proven its reliability.
   d) The layout of the aircraft (size, weight, surface, etc.) was based on the tactical requirements levied against the jet bomber.

III. Comparison of the He 343 with the Ju 287

1. The included table shows a comparison of the weights and flight performance of both prototypes.
2. From this list it is apparent:
   a) The weight of the airframe is:

      For the He 343 = 5,260 kg
      For the Ju 287 = 7,070 kg

      This means that the Ju 287 requires approx. 1.8 tons more of materials to build. In view of the situation with materials, this additional expenditure is only justified if it were to offer a significant improvement in performance over the He 343.

b) The bomb payload for the He 343 is 2,000 kg, the theoretical bomb load of the Ju 287 amounts to 3,000 kg. By carrying a bomb payload of 3,000 kg the takeoff weight of the Ju 287 is brought up to 23,555 kg. With this takeoff weight, the machine can only take off using takeoff assist. Without takeoff assist the takeoff weight for the Ju 287 must, for flight safety reasons requiring a minimum climb angle after liftoff of 3 degrees, be limited to 21.5 tons. There is a particular concern with minimal takeoff reserves when operating with turbojet engines, because as air temperatures rise the takeoff capability decreases for this type of engine compared with piston engines. We saw a significant amount of risk for the Ju 287 under poor takeoff conditions.
For a takeoff weight of 21.5 tons the bomb payload will have to be reduced to 2,000 kg and the fuel to 6,000 kg. This reduction means that the practical usable bomb capacity of both aircraft are virtually the same. Maximum range with the same payload and fuel amounts to about 1,500 km for the Ju 287 compared to 1,600 km for the He 343. The bigger layout of the Ju 287, therefore, offers no advantages with regard to bomb payload and range.

c) For both types, the maximum speed is about the same. According to data provided by the engine manufacturers, the drop in thrust with increase in altitude is less for the BMW engine than for the Jumo engine.
If this data, which our experience has shown not to be the actual case, is indeed applicable, the Ju 287 would be calculated to have a 25 km/h higher speed at 6 km altitude than the He 343.
Data for the Jumo engine, which is included in our figures, has been shown to be quite conservative and has been proven in flight testing.

d) The Ju 287 makes use of 6 BMW 003 turbojet engines. As of yet, this turbojet has not proven itself ready for front-line use. With the unsatisfactory state of development, an increase in thrust cannot be expected.
Use of the better performing He S 011 for mass production is out of the question at this time.

IV. Conclusions from the Comparison Overview

1. The spatial capacity of the Ju 287 offers no real advantages. The He 343 is also capable of reaching the same performance of the Ju 287, but with significantly less use of material and with four instead of six engines.
2. The He 343 can successfully be used as a reconnaissance platform and heavy fighter. This versatility of roles is not on offer with the Ju 287.
3. The He 343's flight testing would offer no new problems whatsoever. It could therefore proceed much more quickly than with the Ju 287, which would bring with it a series of new findings - e.g. forward swept wings - that would have to first be proven
   Mass production of the He 343 could, therefore, be expected sooner than the Ju 287.
4. At the meeting in Berchtesgaden on 25 May 1944 the Ju 287 was discussed favorably, while the He 343 wasn't even mentioned by any of the participants.
   Given the above information, it is shown that the performance of the Ju 287 can be achieved by the He 343 version more cheaply (work hours, fuel) and without taking such a great risk.
   It would, therefore, be in the best interests of our nation's defense to at least issue a contract for full-scale production of the He 343 parallel to the series production of the Ju 287.
   Such a contract would enable us to go back to the stipulation outlined in the initial tasking requirement: to provide the field with a front-line ready, capable, minimal-risk jet bomber as quickly as possible.

Junkers' answer was naturally not long in coming. On 9 July 1944 Professor Dr.-Ing. Heinrich Hertel drafted a biting response to Professor Ernst Heinkel's memorandum:

Statement of Professor Dr.-Ing. Heinrich Hertel
Re: Ju 287 and He 343 Types

Herr Prof. Dr. Heinkel's remarks lead to a blatant fallacy that is basically founded on three fundamental errors in his thinking

1. For the He 343, *Herr* Heinkel calculates the probable output ratios as taken from his official data sheet (No. 1423 from 24 May 1944) using the lower powered Jumo 004 B engine, which has been tested and is currently being mass produced. However, in his comparison with the Ju 287 he uses the improved performance Jumo 004 C engine for the He 343, an engine which has not yet been approved for the STRABO's use. Conversely, *Herr* Heinkel clearly calculates the BMW 003 engine's static thrust at 900 kg in his official data sheet (No. 1423), but reduces this output to 800 kg static thrust for the Ju 287 in his comparison. Accordingly, contrary to an objective, accurate basis for comparison, he has added 115 kg thrust for the Heinkel aircraft and subtracted 100 kg thrust for the Junkers aircraft (as opposed to our own calculations). This difference of 215 kg per aircraft, i.e. a plus of 27 percent, naturally makes the He 343 appear in a favorable light, something which nevertheless is entirely unjustified.
2. *Herr* Heinkel provides data sheets in which he lists the use of various jet engine types in the He 343 model (e.g. No. 1423) and in so doing, overlooks the fact that the Junkers Ju 287 aircraft is obviously capable of and has been designed from the outset for accepting more powerful turbojets, insofar as these are available for production. The use of the BMW 003 for the Ju 287 was presented by the IFA as a first option solely because of the situations with development and acquisition of turbojet engines, but in no event would the use of the Jumo 004 C be ruled out. If *Herr* Heinkel thinks that he can safely use the Jumo 004 C engine with its 1,015 kg static thrust for the He 343, then he must also apply that to the Ju 287 in any comparison, which has taken into account such advanced engines in its overall design layout. When the Jumo 004 C turbojet is applied to both designs, the overwhelming superiority of the Ju 287 becomes readily apparent.
3. In one compilation *Herr* Heinkel estimates data for the Ju 287 that may have corresponded to a project stage at one time or another, but not to that found in the final design stage, meaning that *Herr* Heinkel sets the weight for the Ju 287 too high and its performance too low. For a comparison in mid June 1944 it would have been more accurate to use the weights and performance of the Ju 287 that reflected a current stage of design instead of using some outdated inaccurate documents. Throughout the course of design the RLM has continuously been kept abreast of weight saving measures and the resulting success we've had with adding additional armor and ammunition without exceeding the previously projected weights. Since he contrasts versions that cannot be compared to one another, the weight and flight performance compilation in the memorandum by *Herr* Prof. Dr. Heinkel is fundamentally wrong. The two tables included with this statement should rectify this error.

Table I

Given the state of development and production preparations, the 004 C engine will not be available before mid-1945, and those engines offering better performance output will undoubtedly be channeled toward improving the performance of fighters and heavy fighters under procurement at the time. With this in mind, production startup of a jet bomber using the early B version - i.e. 900 kg thrust - of the Jumo turbojets must be reckoned with. In order to bridge the problems inherent with the fact that the improved performance C engines will not be available for the early startup of the jet bomber, we are initially counting on the BMW 003 engine with the understanding of GL/C-3. In the comparison chart, the engine has been listed with the performance data that has been recommended by the engine supplier as the authorized load, i.e. 900 kp emergency static thrust and 800 kp constant thrust. The table shows that using this estimated comparison the bomb capacity for the Ju 287, at 3,000 kg, is twice as much as for the He 343 at 1,500 kg and the Ju 287's 1,900 km range is 45 percent more than the He 343 at 1,330 km. The calculated speed for the Junkers aircraft is 70 km/h greater despite the additional drag imposed by the gun turret.

Table II

If efforts to make the 004 C engine available earlier and on such a wide basis that it could also benefit the jet bomber are indeed successful, something we should be working towards in any case, then it would of course be used with the Ju 287 as well.

The best objective, indisputable assessment is derived from the comparison of the two types in table II using the same Jumo 004 C engine so highly recommended by *Herr* Heinkel. To obviate any need for discussion, the same data for the He 343 as used by *Herr* Heinkel in his memorandum has been used. This comparison of the two types with the same engine shows the overall superiority of the Ju 287

a) Bomb capacity of the Ju 287 is 3,000 kg instead of 2,000 kg for the He 343.
b) Despite this 50 percent greater payload the maximum range of the Ju 287 is 1,930 km, 20 percent more than the 1,600 km for the He 343.
c) The maximum speed of the Ju 287 is 70 km/h than the He 343 despite the use of a traversable gun turret.
d) As another comparison point, table II shows the Ju 287 with a heavy 4 metric ton bomb load while maintaining the previous fuel capacity. This configuration reveals that the bomb capacity of the Ju 287 is double that of the He 343 for the same range. This is particularly important in that it shows that the bomb bay of the Ju 287 can be used to its maximum capacity (e.g. 2 x 2,000 kg or 8 x 500 kg internally). The He 343 utilizes external racks for increasing its bomb load, which reduces this type's already much lower speed even more. In addition to this vastly superior capability of the Ju 287, the following talking points are also applicable to the type

e) The He 343 does not have a traversable turret despite its significantly inferior speed. Its fixed aft firing guns are considered inadequate. *Herr* Heinkel apparently recognizes the need for a traversable rear armament as well, for in his design blueprints (343.01- 11B from 24 May 1944) he himself offers an option with dual rudders and rear turret as a heavy fighter version. The construction of two entirely different fuselage designs including empennage appears to offer little success. As before, we stand by our viewpoint that it is important to incorporate a gun turret from the outset for such a role and to build it as a complete kit in order to allow installation at any time. Performance data for the Ju 287 takes into account this installation both from a weight as well as a drag standpoint. For the He 343, the fitting of a traversable weapons system involves subtracting the equivalent values from its performance.

f) *Herr* Heinkel's documents do not show whether the He 343 aircraft, contrary to initial figures and the mockup presented in the exchange of ideas, is in fact equipped with a pressurized cockpit, something that is indispensable for a jet bomber. The Ju 287 includes a pressurized cockpit from the outset.

g) The constant references to risks involved with wing sweepback are deceptive insofar as that comprehensive wind tunnel testing has shown that this problem is expected to be remedied in the near future.

Conclusion

As opposed to the points in the "memorandum" that attempt to present a performance parity for the two types using comparisons lacking objectivity, the attached table II shows clearly that the Ju 287 offers significantly better performance when the same type of engine (Jumo 004 C) developmental stage is used. Double the payload at the same range, 70 km higher speeds, traversable defensive armament, pressurized cockpit.

The Ju 287 design work has always taken into account the Jumo engine, and from the outset the basic layout of the Ju 287 has included an increase in takeoff weight to 25 tons necessitated by the use of the Jumo 004 C, meaning that production of the airframe in this version poses no difficulties whatsoever (with the exception of an interruption in startup, since the drawings have now been supplied for BMW engines).

In any event, we can no longer take responsibility for projects that constantly fluctuate back and forth with various assumptions about performance and availability of engines. To eliminate this circumstance of uncertainty on the part of aircraft development and once and for all resolve the issue of juggling engine performance that is unfortunately evident from the other party, I have taken measures to push the installation of both the BMW 003 and the Jumo 004 C developmentally parallel to such a point that installing both these engines in the same manner has been assured. Production, however, needs timely clarification on which type will be fitted, and any reshuffling with regard to previous plans brings a delay in startup.

The remaining points in the memorandum have not been addressed due to time constraints.

Dessau/Raguhn, 9 July 1944

Well, then, both sides certainly have some justification for their position, although Professor Heinkel erred in his data on the Junkers Jumo 004 C jet engine. At this point in time, this engine had by no means proved itself; only the Jumo 004 B with its 900 kg thrust had demonstrated its operational safety. Yet, as always happens when apples are compared with oranges, there remained on both sides nothing but red faces and stacks of paper. Neither one nor the other letter had any serious effect on either development (note: further relevant correspondence on the matter has not been included here, since the two letters above adequately illustrate the problem). On 23 June 1944 the Junkers people suffered their first slight setback at another Fighter Staff meeting:

Participants:
Hauptdienstleiter Dipl.-Ing. **Karl Otto Saur** - Deputy to Armaments Minister Albert Speer
Oberst **Siegfried Knemeyer** - member of the Technical Armaments Staff
Oberst **Edgar Petersen** - Chief of all *Luftwaffe* Test Centers
Oberingenieur **Alpers** - GL/C-B2 Department Head, Aircraft Division, Development Office C, Procurement
Ingenieur **Hübner** - Technical Division Aerodynamics, Design and Flight Testing TEA-FL, at Arado
Dr.-Ing. **Lange** - Junkers aerodynamicist

Alpers: There's to be 50 instead of 100 Ju 287s.

Saur: Was it 50 in the meetings with the *Reichsmarschall*?

Petersen: No. *Generalfeldmarschall* Milch said that at the last meeting.

Saur: I'm in no position to address that. What did the General of the Bomber Forces say to that?

Hübner: It was said that this would happen anyway.

Alpers: In the meeting someone said that the machine was so new that we could easily start out with just 50.

Petersen: No airplane has even flown yet. We would have an airplane that hasn't even flown.

Saur: What does the Ju 287 offer?

Knemeyer: The *Reichsmarschall* has called for a jet bomber. This was a jet bomber design. It was to be started up as a contingency reserve. If we drop it to 50, the companies will lose their entire "pull" ("pull" = advantage) in design and development.

Alpers: The quota has not changed. The companies have until August to deliver 100 machines.

Knemeyer: I stand by my view that we need to increase these 50 units in the interest of the company, even if it's not covered. If we overlook it later, we can still say: it will or it won't happen. Otherwise we'll lose our entire "pull" that's behind it.

Saur: What about the matter of contingency? The 287 is a heavy machine that requires an awful lot of metal.

Lange: If we want to have the 262...

Saur: That has nothing to do with this, *Herr* Lange! We've got to pull ourselves out of this tactical mind set. We don't understand it. We don't need to get involved with this. I can't imagine that we here and now can do without this machine. But we could work it out so that we prescribe: but 50 now and from this point on it's ensured capacity-wise at 100, of which contingency-wise 50 are covered, the program can sustain that. The remainder would have to be covered by redlining other types later on. I would unconditionally make it so the company would have the capacity for the full 100 units, but then 50 from this point here in October. I would make a line here and say: we'll have the machine at capacity with 100, or can be supported capacity-wise at 100, but from a contingency standpoint can only be covered for 50. The increase must be made by redlining other types at the appropriate time in a timely manner. Clear? - Understood? I don't want anyone getting the idea that this machine isn't that important and therefore have it stuck in the testing stage.

Petersen: For now, it's the only bomber we can use against England.

Alpers: I have no reservations. Let's do it this way.

As you can see, *Hauptdienstleiter* Saur tried every trick in the book to keep the Ju 287 alive. The result of this meeting was a confirmation of the resolved facts in the minutes of the meeting on 1 July 1944 that took place with *Reichsmarschall* Hermann Göring. In Section B, paragraph 3 it reads:

Ju 287 100 per month capacity-wise, planned 50 per month material-wise. Further decision on this early 1945.

In the Fighter Staff meeting of 6 July 1944 production startup was defined again. Now, only a total of 21 machines would be built between February 1945 and the middle of 1945, which corresponded to the original contract for 21 machines (three prototypes and 18 pre-production aircraft). It wasn't until the end of that year that a further 50 units would be added. However, things would turn out much differently. The economic restrictions brought on by the increasingly precarious wartime situation for Germany was behind the comment on 18 July 1944 which stated that:

Series production has been ruled out for February 1945.

The 8 August first flight of the Junkers Ju 287 test bed, the Ju 287 V1 (Ju 288 V201), as well as other successful test flights, did nothing to change the fact that this was the beginning of the Ju 287's inevitable demise.

Hitler himself ordered the Ju 287 halted during the meetings of 21 through 23 September 1944. This decision he confirmed in a general report on 25 September 1944, where under heading 7/925/1245 it reads:

In conjunction with concentrating all energy on the most urgent tasking the *Führer* has agreed that all work on the Me 264 and Ju 287 types, to include test aircraft, is to be canceled effective immediately.

Everything had started off so well, and even in early September 1944 the folks at Junkers were especially optimistic about the Ju 287. An overview on the state of development was compiled that included a status report effective 15 September 1944 with a description of the individual test aircraft as well as a construction description. With the cessation of production the Junkers company felt compelled to seek reimbursement for its costs up to that point working on the STRABO. In the Aircraft Operations Information Report for the month of September, dated 10 October 1944, with regard to the Ju 287 project it reads:

On the orders of the Armaments Staff work on V1-Vv6 was canceled in late September/early October. In July the offer for development and construction of V1-V2 (flying mockups) was assessed at a value of TRM 4,120. The costs associated with the additional prototypes and series production (approx> TRM 6,000) must still be tendered.

Aside from the costs, which seem to leap out, this information report casually mentioned that by this time work had already begun on the V3 through V 6 prototypes in addition to the already com-

Perspective sketch of the first Junkers Ju 287 V1 testbed.

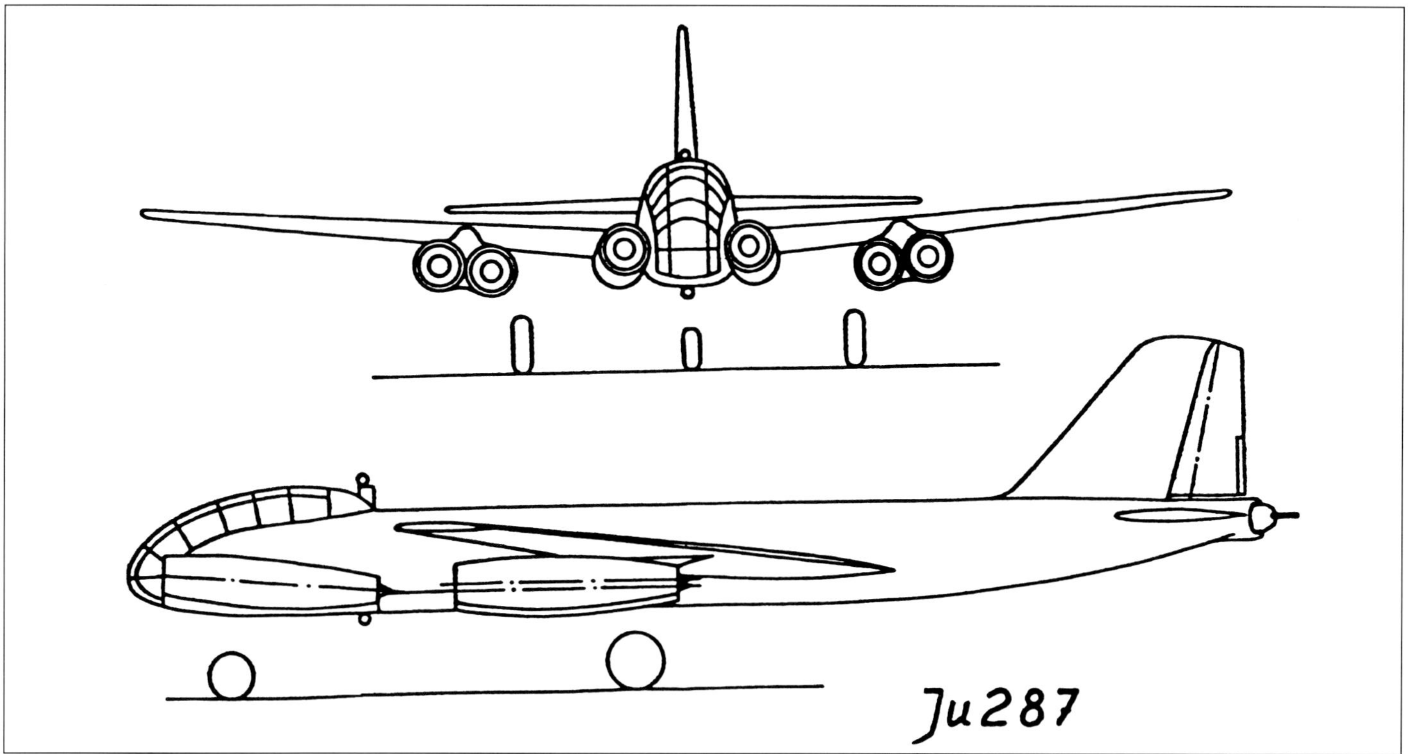

Design sketch of a six-engined variant with periscope sight and tail guns that never made it off the drawing boards.

pleted V1 and V2 prototypes, which were identified here as flying mockups. In early 1945 the SS began to take an increased interest in the Ju 287. They needed a superior strike weapon and saw in the Ju 287 a suitable weapon for their own unit of flying forces they envisioned, the *Flieger-SS*. So it is not surprising to find *SS-Standartenführer* Kloth participating in the meeting with *Hauptdienstleiter* Saur on 10 January. Since, however, the RLM had lost interest in the Ju 287 for the time being, the Junkers company attempted to attract the interest of the Japanese for their high tech product. On 2 February there began the first license negotiations with the Japanese, with the whole thing running under the designation: Japanese Army License Negotiations - Copy Tasking "N" ("N" stood for Nippon = Japan). The negotiating team for the Japanese consisted of General Otani, Major Kobayashi, Major Susuki, and Major Kawakita. The result of these negotiations was

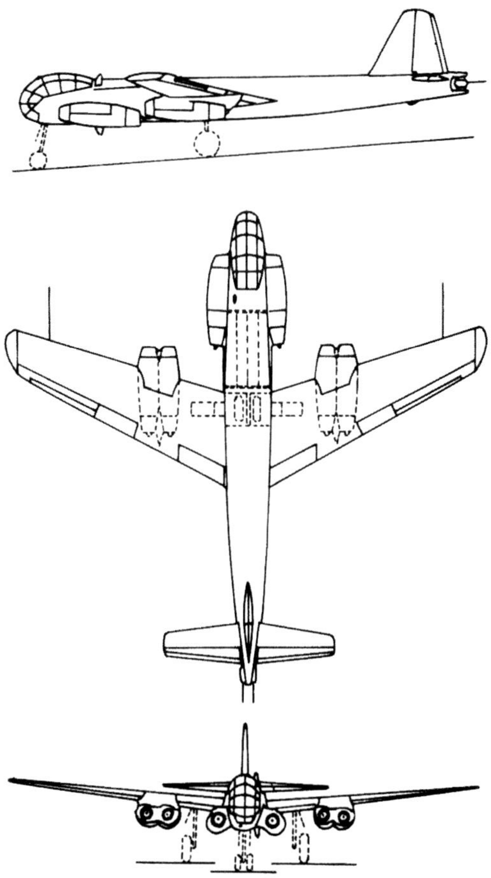

Three-view of the Ju 287 production version as it appeared in early publications; however, this too was never realized.

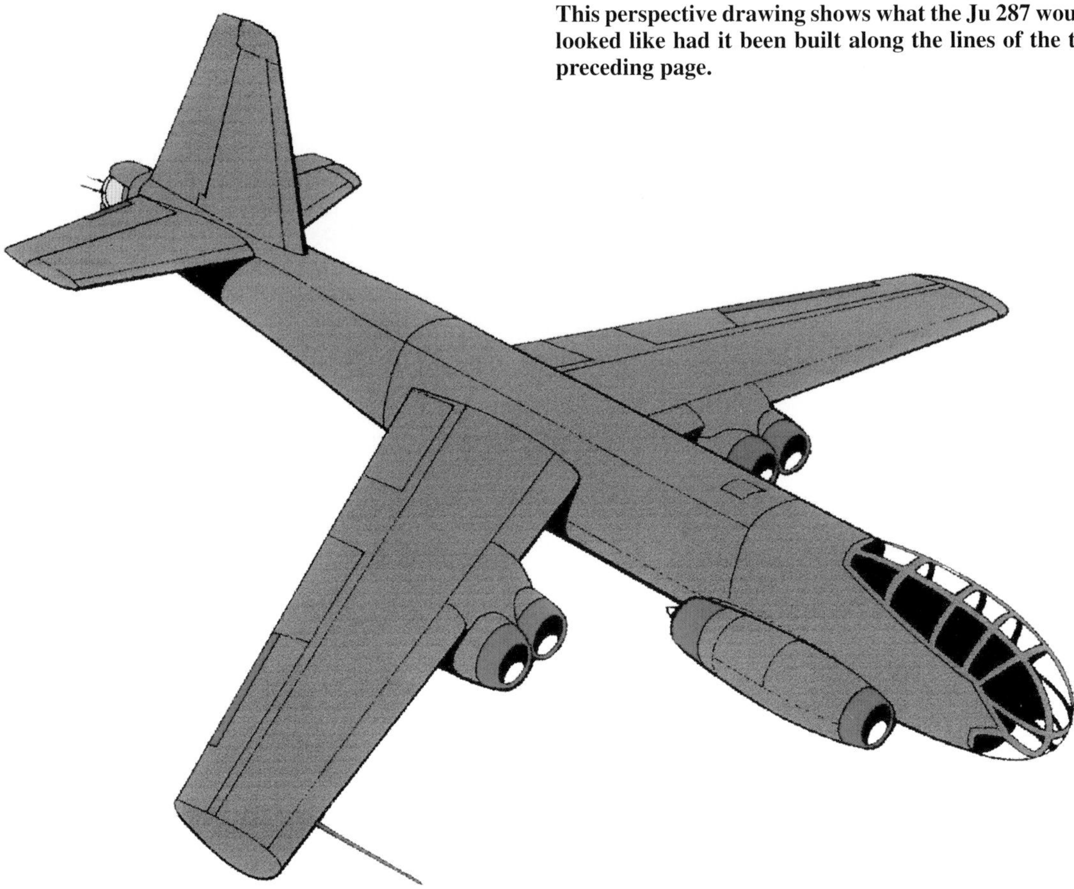

This perspective drawing shows what the Ju 287 would probably have looked like had it been built along the lines of the three-view on the preceding page.

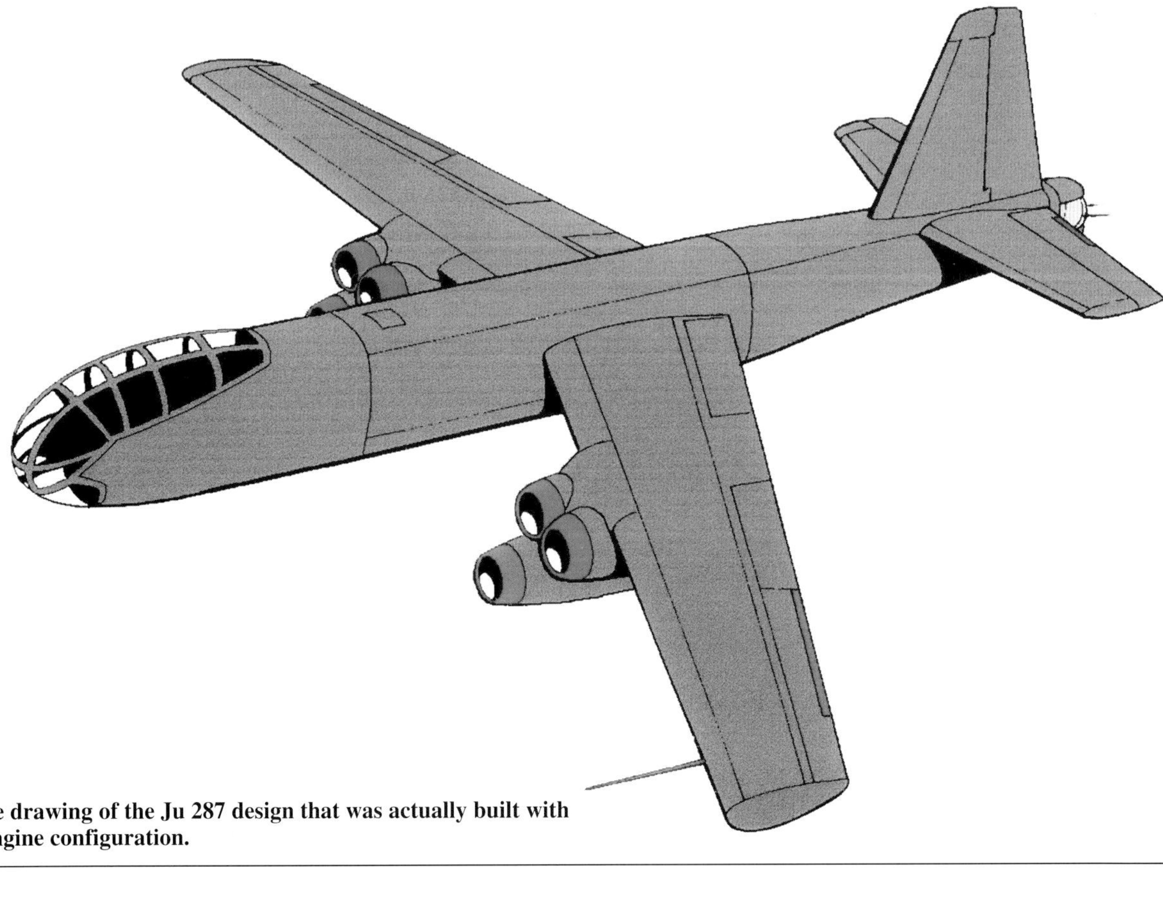

Perspective drawing of the Ju 287 design that was actually built with its triple engine configuration.

the delivery of two sets of blueprints, which had been fully compiled and packed in crates by 3 February 1945. However, before these documents could make their way to Japan, according to records from 27 March 1945 they were destroyed. Item 4, Ju 287 notes:

> Two sets of documents were compiled in full according to memorandum FTH-FN 14/15 from 2/3/45 and partially packed in crates. Relevant delivery listing did not exist; the documents were completely destroyed.

What had happened? An air raid destroyed a large percentage of the documentation for the Japanese at Junkers, including the complete blueprints for the Ju 287. The memorandum states succinctly: "As a result of enemy action...the documents were destroyed." And at the end it says "The reproduction of the documents in their previous scope is hopeless." Why this? Now, you must understand that in the meantime there was renewed interest in bringing the Ju 287 back to life and as a result there could no longer be any talk of passing this top quality product of German engineering on to the Japanese. Indeed, there were even rumors that an actual bombing raid on the Junkers plant was used as an alibi for no longer providing any documentation to the Japanese. But we're getting ahead of ourselves.

As stated earlier, there were efforts on the part of the SS to assume the Ju 287 project and this led to the Ju 287 once again being placed into the spotlight of desire in early March 1945.

On 14 March there was this entry in the war diary:

> Ju 287 is to be immediately addressed as a top priority action, under the direction of H. Tiedemann. Goal is the engagement of shipping targets at long range. Additionally, the optimum solution is to be tackled with every effort.

At another point it says

> Aircraft again placed under technical production care. V3 prototype, V4 prototype for full-scale production. Contract for manufacture of 75 units.

A meeting with the General Staff, General Quartermaster Division 6 (*Genst. Gen. Qu. 6*); Dept. with Command Staff and Weapons General, from 17 March 1945:

> In the first stage the Ju 287 will be built, in the second stage the Ju 287 with increased fuel quantities using RATO. Determination of a new project has not yet been addressed.

Now there was suddenly an effort to make up for lost time. The progress of the war, however, did not even allow for this modest production run. By the war's end not a single production aircraft had left the Junkers Works assembly line, let alone fly a combat mission. Only the V 3 prototype was in an advanced stage of construction, and component assemblies for some of the production aircraft had been built. Thus ended Priority Action 287, before it had really every begun, in the chaos of the final weeks of the war.

But those in the design bureaus still had their eyes on other projects based on the Ju 287, such as the long range version under the designation of Junkers Ju 287 S. In order to attain a range of 4,550 km at an altitude of 11,300 m it was necessary to make use of the He S 011 turbojet engine. To carry the fuel needed, the takeoff weight was increased by 8,200 kg, thereby doubling the fuel supply - an impressive feat.

Other projects envisioned the use of the Ju 287 as the explosive part of a superheavy *Mistel* team, the so-called "*Rückstossmistel*" (literally: Recoil Mistletoe). This involved using a twin-jet Me 262 or, as a backup, a single-engined He 162 as the control plane. The entire combination was to be launched into the air using a takeoff cart upon which rested the Ju 287 filled with explosives with the Me 262 on top. Takeoff boost would be provided by additional Walter 109-502 D RATO packs attached to the cart which would help to heave the beast into the air. The takeoff trolley would detach immediately after takeoff and come to a halt with the aid of a braking 'chute, much like that of the Ar 234 A version. There was even a data sheet drawn up for this version, called the *Mistel* 5 (note: in several documents the same designation was also applied to a combination consisting of the Ar 377 with an He 162 control plane. This is understandable in view of the fact that the *Rückstossmistel* was called the *Mistel* 5 without having determined the final combination of explosive platform and control plane).

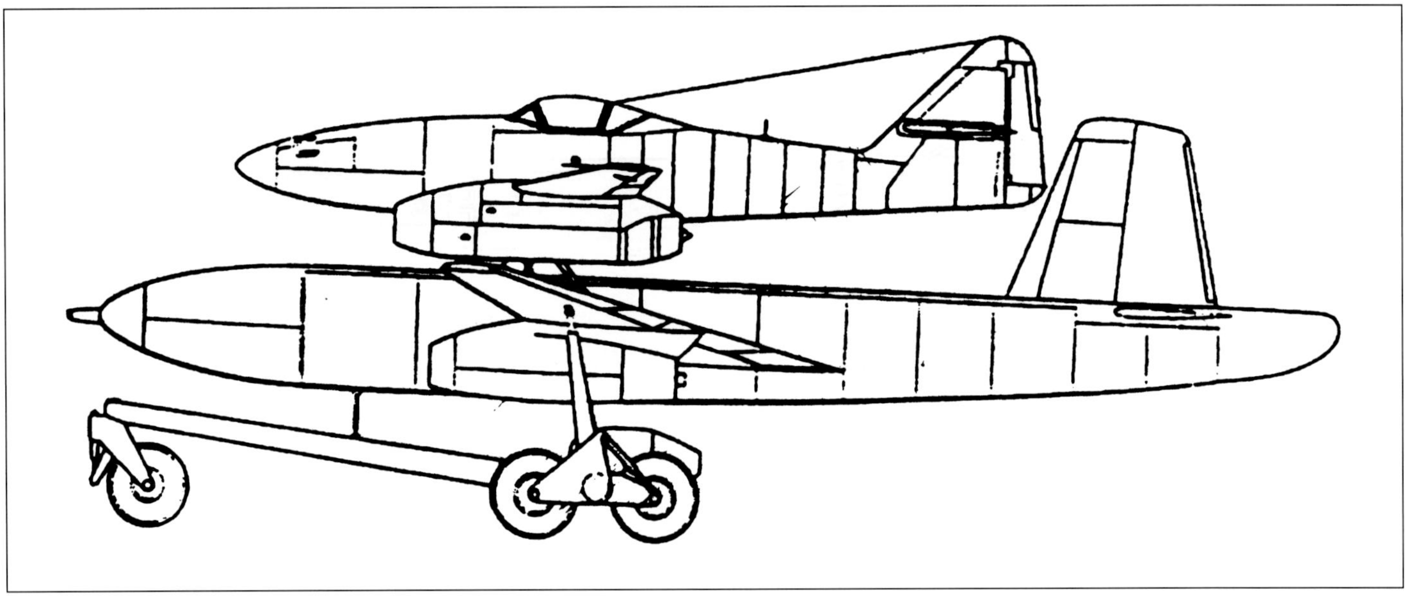

Project study for a "*Rückstossmistel*" piggyback team consisting of an Me 262 controller plane and a Ju 287 bomb platform.

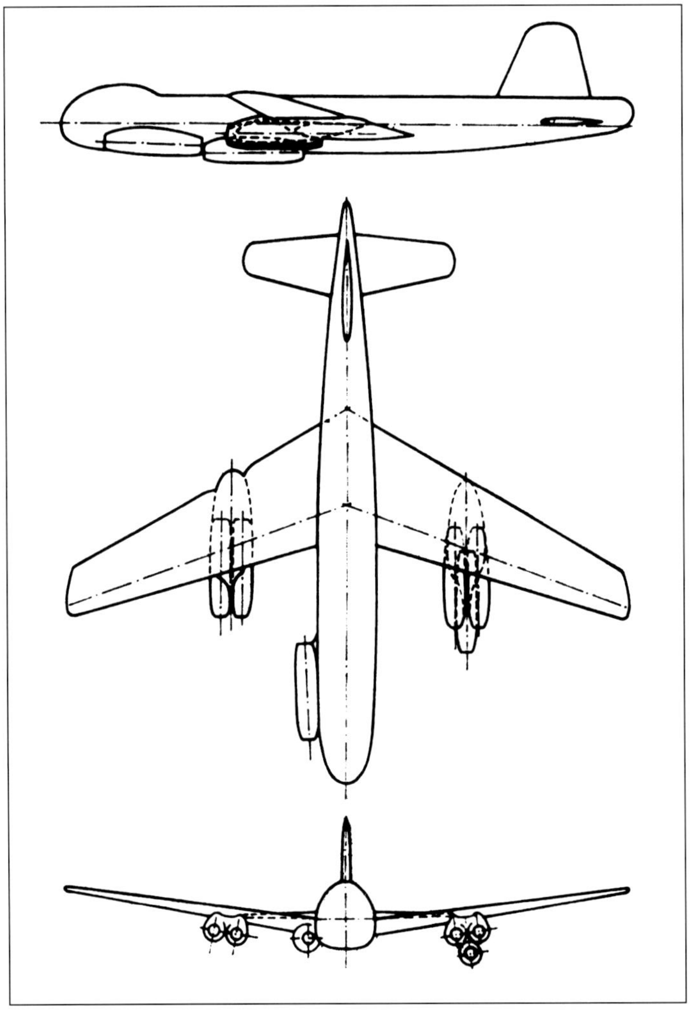

Three-view sketch of the Ju 287 S with different engine pod layouts.

Another project saw the Ju 287 laid out as a long range reconnaissance platform. In this role the Ju 287 was to be fitted with three RB 75/30 cameras in the bomb bay, with one camera positioned vertically and the other two angled 20 degrees to the right and left, respectively, in order to provide the widest area of coverage possible. The remaining space in the bomb bay was taken up by a 2,000 liter auxiliary fuel tank. In order to save weight the metal bomb bay doors were removed and replaced by wooden panels. Yet another version was envisioned as a night reconnaissance aircraft fitted with two RB 35/25 cameras in the bomb bay. It was planned to carry 16 flares in order to illuminate the area to be photographed. Again, the remaining space in the bomb bay was used for a 1,200 liter´tank. The night reconnaissance platform was to retain the bomb bay doors in order to provide a clear field of view for the cameras while at the same time being able to release the flares.

Nevertheless, none of these projects was ever realized. With the American occupation the Junkers Works on 16 April 1945 and its subsequent handover to the Soviets on 1 July 1945 the history of the Ju 287 came to its conclusion - for the time being.

The next chapter will take a detailed look at the Junkers Ju 287 V1 (Ju 288 V201) test bed and its flight testing program.

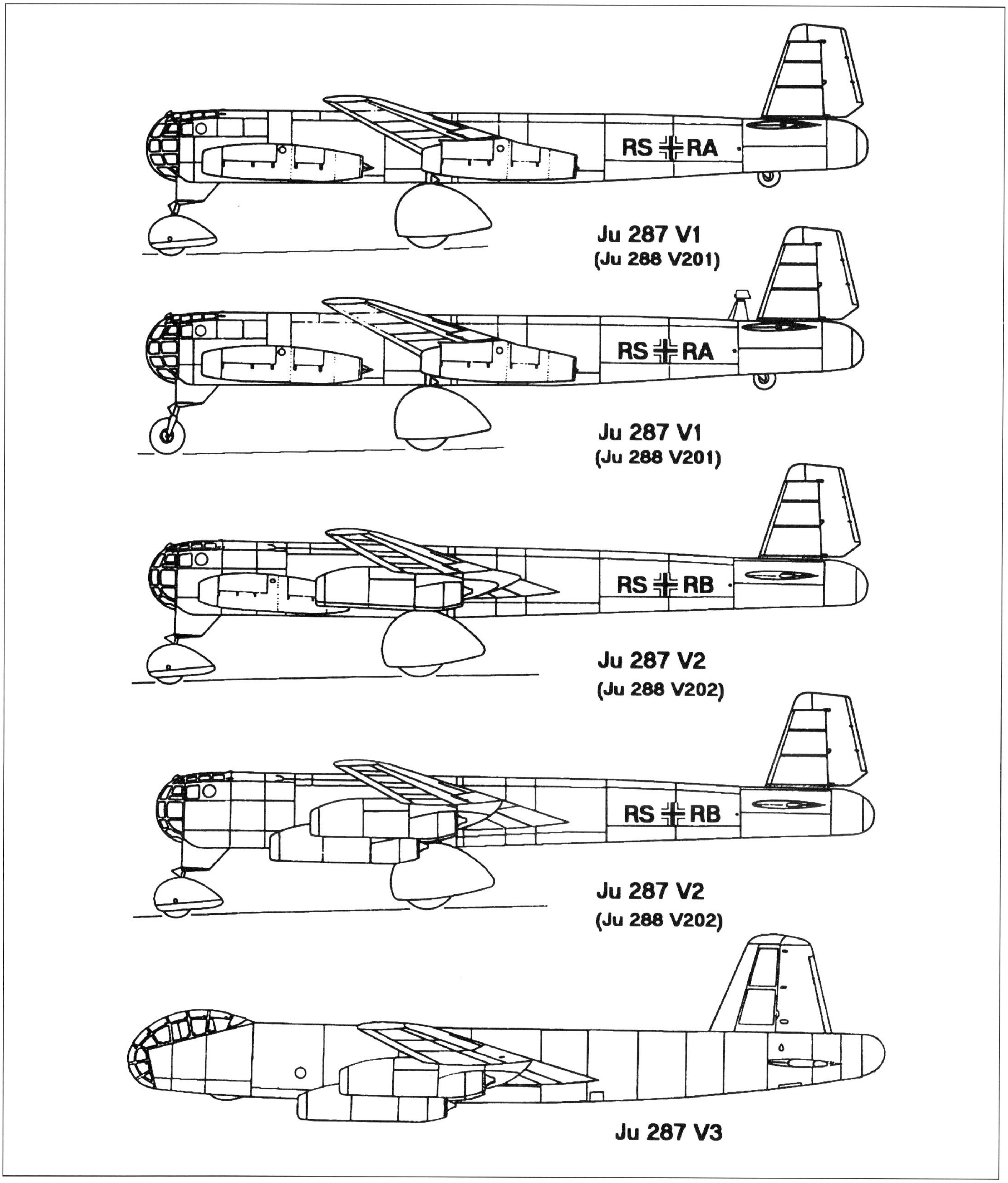

Evolutionary chain of the Ju 287, top to bottom: Ju 287 V1 with nosewheel spats, Ju 287 V1 with spats removed and camera mounted ahead of the rudder for capturing on film the airflow characteristics in the area of the wing root. Airflow was charted using wool tufts glued onto the skin. The next view is of the Ju 287 V2 with the original engine configuration, while below that is the V2 with the triple engine layout as it ultimately was tested. The final side view shows the Ju 287 V3 as it was to have been delivered minus its periscope and tail armament.

Testing the Junkers Ju 287 V1

Since the comprehensive wind tunnel tests could only confirm theoretical findings at best, one could not be absolutely certain how the new wing would function in practice - seldom do theory and practice lead to an ideal marriage - and it wasn't long before the general consensus was to build a simple test bed in order to find out for certain.

The Ju 287 V1 (Ju 288 V201) already described was a composite aircraft built from existing components taken from a wide variety of airframes. Among other things, the fuselage of an He 177 was selected because its dimensions were quite similar to that of the later Ju 287 production version. Another unique feature was the Ju 287's use of ejection seats for the test crew. In order to

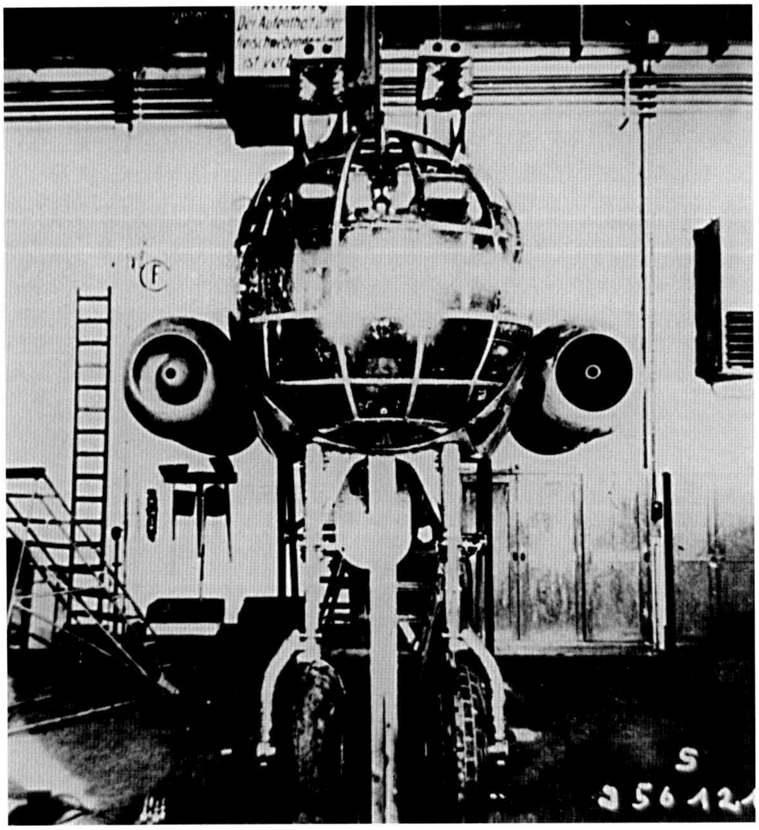

Front view of the fuselage nose mockup of an He 177 with cheek mounted JUMO 109-004 B jet engines.

accommodate these ejection seats the standard cockpit of the He 177 had to be modified accordingly. Particularly, changes to the area around the cockpit roof were needed. The entire roof had to be built so that it could be jettisoned, as this needed to separate before the ejection seats could be fired. Separation of the cockpit roof was initiated by a compressed spring that was activated by the crew in the event of an emergency; only then could the ejection seats be used.

Like the Heinkel company, Junkers had carried out experiments with ejection seats. These ejection seat experiments were chiefly conducted from a Ju 87, the famous Stuka dive bomber. For these trials, the aft canopy on the Ju 87 was removed. Initially using a mannequin as a guinea pig, the ejection seat was successfully tested at increasingly greater dive speeds. Later the well-known parachutist Wilhelm Buss volunteered himself for the experiments and made several successful ejections using the seat (Buss was the first person in the world to successfully eject from an aircraft, using an ejection seat from an He 219).

The individual components of the Ju 287 V1 were shipped to the airfield at Brandis between 21 and 27 May 1944, with final assembly taking place there in Hangar IV. On 28 May 1944 Brandis suffered a major bombing raid at the hands of the US Air Force. Amazingly, the Ju 287 V1 survived the attack unscathed.

The Ju 287 V1 was fitted with aluminum wings, whereas the production version was intended to use steel for the spars and skin. It was powered by four Jumo 004 B turbojet engines with a combined static thrust of 3,560 kp. Two of the engines were mounted on either side of the fuselage, just behind the cockpit, and the other two engines were situated about one-third the way along the trailing edge of each wing.

Above right: Side view of the He 177 nose mockup with the fuselage mounted engines. Above: Final assembly stage of the first Junkers Ju 287 V1 testbed in the assembly hangar at the Brandis airfield. The fuselage engines are still on their blocks, waiting to be mounted.

Left: The same assembly stage taken from above showing the negatively swept wing layout. (Photo: *Deutsches Museum*)

Because it was such a pretty sight, here is another view of the same situation, taken this time from above and behind. This photo shows the Ju 287 V1 barely fitting into the hangar and how at times the hangar doors had to be opened for certain types of work. (Photo: *Deutsches Museum*)

After final assembly had been completed, the long awaited first flight took place on 8 August. There had been repeated delays because urgently needed parts had not been delivered. The wait for the engines was also a long one; they were needed just as urgently elsewhere. Even before 8 August taxi trials had been carried out at the gigantic airfield complex at Brandis. Positioning the aircraft for takeoff proved to be problematic, as the dual nose wheels were not steerable. A jury-rigged steering mechanism utilized two cranked rods stuck into each side of the dual nose gear, with one or two men on each rod to set the position of the nose wheels and/or guide them during slow taxiing using nothing but pure manpower. A somewhat awkward procedure, but one that in no way impacted upon testing operations. Once set in the proper position, the wheels were force guided on takeoff as well as during landing. A standard method of steering even today for some types of single-engine sport planes with nose wheels. However, with a hefty push on one of the wheel brakes these 'lightweights' are much easier to manage, as the nose wheel immediately shifts in the right direction. Incidentally, experience had already been gained using the rod control with the engine-less Me 321 *Gigant* (Giant), a transport glider with a massive wingspan of 55 m, the second largest that has ever been built. It was only surpassed by the Junkers Ju 322 *Mammut* with the monstrous span of 82.35 m (many sources note the more realistic wingspan of 'just' 61.70 m or 62.35 m). The two non-steerable nose wheels of the Me 321 were also positioned prior to takeoff by ground crew using rods and not a little elbow grease (see pictures above and below).

This detail shot shows the fuselage mounted engine pods to good effect as well as the work on the ejection seat installation. Fitting the ejection seat necessitated removing the entire upper cockpit glazing.

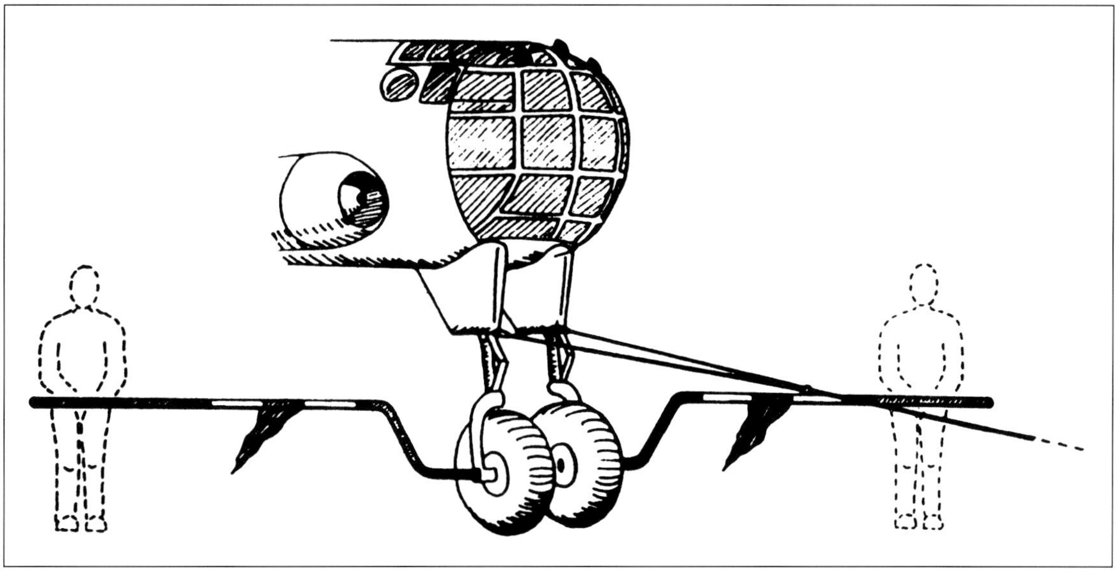

This drawing by Mr. Uwe W. Jack (Aviatik Vol 2) shows his correct interpretation of steering the Junkers Ju 287 testbeds while taxiing.

During the day of 8 August the heavy machine was rolled out of the hangar and made ready for its maiden flight. To shorten the takeoff run three Walter HWK 109-501 A rocket packs, each with 1,000 kp static thrust, were attached. In actual fact, four of these units were to have been fitted, with two being ignited for the takeoff and the two remaining RATO pods acting as a backup in the event that one of the pods used for takeoff failed. After a brief discussion it was determined, however, to only use three pods, with two of these fired together on takeoff as planned. The third one would be ignited just after takeoff to give an added boost on climb out. This became the common practice; if one of the packs should fail on takeoff, then there was always the third one available for emergencies. Nevertheless, on the plane's first flight all three were ignited - apparently an oversight, but perhaps to ensure enough thrust for a safe takeoff. But there was indeed an unanticipated problem, as the report from Walter Rebaum, an eyewitness to the maiden flight, clearly reveals. More of this anon. There is no photographic evidence available that verifies that four Walter RATO packs were ever used for takeoff; only three such assisted takeoff pods are ever visible, leading one to assume that only three Walter RATO packs were ever used.

This photo of an Me 321 "*Gigant*" shows that people knew of the necessity of manually controlling a dual nose wheel configuration using "shaft steering" long before the Ju 287 V1. Here, too, rods had to be used prior to takeoff in order to properly line up the dual undercarriage (which in this case was jettisoned after takeoff). To facilitate this, each wheel was equipped with a hollow shaft, with a second opening for rod insertion located above the wheels (as seen in the photo). This served to activate the brakes so that the tow plane(s) could pull the tow cable taut before takeoff.

But let us return to 8 August 1944. Another eyewitness to the maiden flight carried out by *Flugkapitän* Siegfried Holzbaur was the Junkers employee *Dipl.-Ing.* Fritz Freytag. Here follows his short narrative:

> The 8th of August was a sunny day. In the early AM the machine was rolled out of Hangar IV and readied for takeoff by the ground crew. *Flugkapitän* Siegfried Holzbaur had prepared himself well for this first flight. In order to gain experience with operating jet engines he had carried out several takeoffs with a jet-powered Me 262. Then, finally the turbojets were started up. The heavy machine slowly began to move. As it rolled, all three Walter takeoff assists were ignited. In a huge billow of smoke the Ju 287 leaped forward and accelerated visibly. The heavy smoke prevented us, who were some way away from the events, from seeing how the heat from the combined exhaust gases was so powerful that parts of the concrete runway were broken up and blown away. We watched spellbound as the machine growled along. Ponderously the airplane lifted itself into the air and, despite the help of the assisted takeoff pods climbed away at a very shallow angle, disappearing from our view shortly afterward. We couldn't see the machine any more, but there was no mistaking the sound as it flew off. After awhile the airplane came into view again at an altitude of about 2,000 m, overflew the airfield and jettisoned the RATO pods some distance away, which fell to earth drifting on their parachutes. The heavy machine droned over the airfield a few more times, then touched down for landing in a completely unspectacular fashion and taxied in front of the hangar. As *Flugkapitän* Siegfried Holzbaur later reported, there were no problems with this first test flight of the Junkers Ju 287, during which the airspeeds never exceeded 380 km/h.

Other than the concrete damage to the surface of the runway, which is repeatedly mentioned but has never been confirmed in documentation, *Herr* Freytag reports that no other problems were encountered. This is entirely different from the very detailed report by *Herr* Walter Rebaum filed somewhat later, which shows that not everything went as smoothly as mentioned and that there was a whole series of complaints. There followed other, equally successful test flights. During the course of the experimental flights it became apparent that the theoretical calculations and wind tunnel testing results in large part were in harmony with the results from the flight tests. There were constant problems with the engines, and once or twice one of the Walter pods exploded, but none of this affected the tests. It turned out that, as discovered in the wind tunnel tests, the two fuselage mounted engines acted as canards and ensured that the airflow did not separate even on approach at lower speeds. This meant that the short slats attached to the inner wing sections, which were designed to automatically extend at low speed when the airflow separated, could be dispensed with. As a result, these slats were deactivated and quickly patched over. All in all, the forward swept wing demonstrated a poor lift coefficient, a peculiarity that had already been noted from the wind tunnel tests. To correct this bad habit an additional tiny slat was fitted to each of the wing center sections.

In addition, the less attractive features of the negative swept wing increasingly became evident. This tended to twist during tight turns at high speeds. Junkers had recognized this problem in advance and as a countermeasure had given the wing profile a twist at the wingtip in the opposite direction of the force, so that the negative effects could be kept within acceptable limits as the wing flexed when overloaded.

Takeoffs were made both with and without takeoff assistance in order to determine the takeoff run with and without RATO. In order to provide a better picture of the airflow behavior on the new wing, someone came up with the idea of glueing wool tufts to the wings and parts of the fuselage. Walter Rebaum, in the rank of

This detail photograph of the undercarriage shows one of the inserted "guide rods" used for taxiing in the lower left corner of the picture.

This photo gives a good impression of the size of the nose undercarriage spats in comparison with a human. It looks like the man is fitting one of the "guide rods" to the nose gear.

Gefreiter (lance corporal or airman equivalent) at the time and his assistant Marianne from Brandis were ordered to glue these tufts to the wings and fuselage/wing joint area (see photo page 62). To constantly monitor the airflow characteristics during flight an automatic camera was installed just ahead of the rudder that would record the movement of the wool tufts throughout the flight. Thus it was possible to assess the airflow characteristics on the wing itself and at the wing root following the completed test flight. Walter Rebaum lives today as a retired machine engineer in Hamburg and at the time often had the opportunity to observe the Ju 287 during its test flights:

> Beginning in the spring of 1944 the Junkers company from Dessau was given the use of a hangar. As soldiers, we had no knowledge of what took place there. We were only aware of what we saw or heard. You were deliberately kept in the dark, if only for reasons of secrecy according to the maxim: "the enemy is listening".
>
> Thus, Junkers only had a Junkers Ju 88 parked outside the hangar. A cylindrical 'container' was attached beneath the fuselage of this aircraft, bout 60 cm in diameter and three meters long. We believed that this might be an auxiliary fuel tank. Then I learned that a flight instructor related how this very machine had overtaken his Ju 88 in flight, with its engines off and with its props feathered. Then it became clear to me that this must be a turbojet engine, something I'd already heard rumors about.
>
> I think that it was around the end of June 1944 when I was detached from my technical company to Junkers to work there.
>
> Naturally I was quite surprised to discover a fully complete airplane there in the hangar. In the technical company we had to work 12-hr shifts, and none of us noticed that the Junkers company had brought in parts for a complete airplane.
>
> I was assigned to the group that was responsible for takeoff preparations. We installed a braking 'chute in the tail of the machine that could be deployed from the cockpit by means of a Bowden control cable. The aircraft was fueled up, and a tractor towed it to the calibration hard stand. There it was anchored down and the tractive forces measured with engines running using a dynamometer. It was then that I heard for the first time in my short life the howling sound made by a running turbojet engine. The machine was fitted with four Junkers 004 engines. Each one had a thrust of 900 kg. Peanuts compared with the performance of today's engines. But at the time it was for us a technological revolution! And I was there! This plane had two nose wheels. There was no previous experience with how the plane would behave when taxiing on the ground, so the machine was towed by tractor to the runway and the wheels were set in the proper direction using two rods.
>
> A checkerboard pattern was painted onto the outside of the fuselage. This was thermopaint, with which you could tell how warm the external skin became from the exhaust of the turbine engines. The Americans and British were unaware that we in Germany were making advances in this arena, nor did they know that we were already installing ejection seats. Then I later learned that Heinkel, not Junkers, built the first ejection seats.
>
> A technician was present when the first Walter RATO packs were fueled. We knew of the dangers with *C-Stoff* and *T-Stoff*, but there wasn't the requirement for protective gear that we have today. There were only protective goggles and rubber gloves.
>
> I remember that there was good weather on the day of the first flight. That day we mechanics had to stand at 100 meter intervals along the edge of the runway to act as markers for measuring speed/acceleration.
>
> A Ju 188 took off alongside the Ju 287; it was to have filmed the takeoff. Everything went quite well, even the three RATO packs ignited. But when they'd burnt out, the machine pancaked dramatically, but was able to pull up. For later flights the RATO pods were ignited at timed intervals.
>
> After about 40 minutes the machine landed again and rolled alone over to the hangar. The braking 'chute wasn't needed.
>
> A lot of prominent people had come; naturally we held them back. You should have heard all the praise about the successful flight! For our part we got several pages of paper with complaints that would need correcting. - The spats on the fixed nose wheels fluttered during the flight;

we therefore later took them off. A film camera had been installed ahead of the rudder, which was used to film the strings that "Marianne from Brandis" and I had glued to the wings. We wanted to know the conditions under which the airflow could separate from the wings.

During subsequent test flights the takeoff length was measured with and without assisted takeoff, as well as the length of the landing run with and without braking 'chute. Further successful experiments with restarting the engines in the air were conducted. Rolls were carried out in the air, attempts at stalling, etc...

There were probably ten or twelve test flights that took place. Our group took care of preflight preparations, fueling, etc. When they had burned out, the RATO pods were jettisoned and came down on parachutes. We recovered them. This was done very carefully due to the dangerous fuel residue, which could potentially reignite. Later the parachutes were refolded and reattached.

In the meantime, the second test bed had arrived at Brandis. It was fully assembled and had six jet engines. Then we heard that everything was stopped, apparently because the fuel was being switched from leaded benzine to kerosene.

Without any explanation, in late September 1944 I was ordered to Basdorf near Berlin, to the BMW engine plant there. There, still a soldier in the *Luftwaffe,* I worked in the engine testing workshop. The problem was that there was no heat-resistant plates for the gas exhaust jets. We experimented with black iron plate. After a short time it was burned up and splattered on the wall. Then we worked on different airflow divergence designs like those that can be found today with modern engines. We soon were able to log 25 hrs service life...

Detail view of the checkerboard-like thermopaint behind the starboard engine, designed to measure the skin temperature caused by the turbine exhaust gases.

Once the Ju 287 V1 has been towed into position, it is fueled and checked over. These are probably last minute preparations prior to its maiden flight.

The front view shows to good effect the lay-out of the four engines. The engine intake covers have already been removed. The crew's entry hatch is still down and the boarding ladder still extended.

Front view of the Ju 287 V1 prior to the rocket-assisted takeoff packs being attached, seen just before its first flight on 8 August 1944.

Large-size photo of the front nose section of the Ju 287 V1. Despite some of the engine panels being removed for maintenance, it is easy to recognize the external lift inducing shape of the cowlings with their extended length and a cross section similar to that of a wing.

Above: Nice shot taken shortly after rollout from Hangar IV. A parked Ju 88 can be seen in front of the hangar doors in the background to the left.

Right: Final preparations for the first test flight on 8 August 1944, just before its maiden flight. One last time the technicians check everything down to the finest detail.

The photo to the right shows Walter Rebaum in 1944 as a technical soldier in the *Luftwaffe*. Walter Rebaum, born on 8/14/1924 in Arnis (Schleswig) was called up for service in the *Luftwaffe* on 10/20/1943. There he completed his training as an aircraft mechanic at *Flugzeugführerschule A/B 33* at Altenburg/Thuringia. Following that, he was retrained as an aircraft engine mechanic at the *Fliegertechnische Schule* in Detmold starting on 04/30/1943. On 7/7/1943 *Herr* Rebaum was posted to *Blindflugschule 1's* technical company at Waldpolenz near Brandis, from where he was assigned to the Junkers company at the Brandis airfield from June 1944 onward. There he worked as part of the technical group responsible for test flights. At the war's end *Herr* Rebaum was made a French prisoner of war and was lodged with a farmer's family. A happy ending, for he still has friendly ties with this family even today. Once he was released and returned home, *Herr* Rebaum studied constructional engineering at the *Ingenierschule Hamburg*, from where he graduated as an engineer in 1952. After a fulfilling life of work he looks back with interest on the events of that period from his well-deserved retirement today.

This well-rendered drawing by Gert W. Heumann shows the first flight of the Ju 287 V1 at Brandis.

Walter Rebaum and his assistant Marianne hard at work glueing the wool tufts to the wings and fuselage at the wing root.

Frames from a film showing the maiden flight of the Ju 287 V1 on 8 August 1944. Despite the extremely poor quality of the film, it is still possible to make out the Ju 188 taking off parallel to the accelerating Ju 287 V1 as mentioned in *Herr* Rebaum's report. The Ju 188's tailwheel has already lifted off the ground, but it is clearly evident that the 188 is having trouble keeping up with the faster Ju 287 V1 and is already falling behind somewhat. The second frame shows the Ju 287 V1 lifting off with its Walter RATO packs still trailing smoke. Other images from this film sequence exist, but these were not available.

These two photos show quite clearly just what Walter Rebaum and his assistant did. The glued on wool tufts are easily discernable. The film camera can be seen just ahead of the vertical stabilizer. The RATO pods have been attached, indicating an imminent test flight. This photograph is probably the most well known of the Ju 287 V1. (Photo: *Deutsches Museum*)

This highly interesting report by Walter Rebaum contains many details that might otherwise be quite easily overlooked. From this, we now know that all photos of the first test bed with the wheel spats affixed were taken before the first test flight, and all other photos were taken afterward.

The thermopaint mentioned by *Herr* Rebaum was a type of paint that, with the addition of chemicals (copper, cobalt, nickel, chrome, molybdenum, chlorides, etc.), would change color irreversibly with the shades based on the temperature reached. A method that, even today, offers a simple means for measuring the temperature of cast iron in smelters for example, where you mark the iron with a thermocolor pen and wait for the color to change to the last point at which the iron remains stable. But now let's return to the testing of the Ju 287.

On the other hand this photograph, very similar to the previous one but taken from a different viewing angle, is probably the least known photo of the Ju 287 V1. It shows several details not seen in the earlier photo, such as the RATO pod beneath the right nose engine as well as the canopy's break lines on the upper fuselage for egress using the ejection seat. (Photo: *Deutsches Museum*)

26 September 1944 - lots of activity at Insterburg. *Flugkapitän* Pancherz (on the far right with light colored jacket and cap), shortly before the enormous Ju 390 V1's demonstration flight for Adolf Hitler and his staff.

In addition to *Flugkapitän* Siegfried Holzbaur the Ju 287 V1 was also flown by *Flugkapitän* Hans J. Pancherz, whose flight log we still have. Pancherz was not an unfamiliar name when it came to test pilots for the big Junkers designs: on 20 October 1943 he flew the six-engined Ju 390 V1 at Merseburg on its maiden flight and on 26 November 1943 flew the Ju 390 V1 before Hitler and his staff at Insterburg.

Flugkapitän Pancherz flew the Ju 287 V1 with the markings of RS+RA on 8 September 1944 from 0829 hrs to 0953 hrs, i.e. 84 minutes long, before landing at Brandis again. Also on board was *Flugkapitän* Siegfried Holzbaur, who was there to instruct him in handling the Ju 287 V1. On 13 September 1944 Pancherz flew the Ju 287 V1 exactly two times. He took off, again from Brandis, at 0907 hrs, only to land after 38 minutes at Brandis at 0945 hrs. That afternoon he boarded the Ju 287 V1 yet again, took off at 1735 hrs and after 55 minutes of flying time landed again at 1830 hrs without incident. All three test flights concluded without any notable dramatic occurrences whatsoever. Unfortunately, there is no further documentation available that shows any pilot other than *Flugkapitän* Pancherz and *Flugkapitän* Holzbaur flew the Ju 287 V1, as has repeatedly been claimed.

In all, a total of approximately 15 to 20 test flights were carried out, during which a speed of 680 km/h was reached without any serious problems being noted. The Ju 287 V1 was transferred to the Rechlin Test Center for follow-on evaluation in the fall of 1944, and there it was photographed by an Allied reconnaissance plane on 31 September 1944 - an event that has caused some confusion because someone typed in a 'five' for the month of May instead of a 'nine' for the month of September.

Front view of the Ju 287 V1 shows how it appeared in the later stages of testing with its nosewheel spats removed. The RATO pods have been fitted for the impending test flight. It is obvious that only three pods have been mounted.

Final takeoff preparations before a test flight with the Ju 287 V1. *Flugkapitän* Pancherz (wearing the cap) is seen talking with one of the maintenance personnel. As *Flugkapitän* Pancherz only flew the Ju 287 V1 a total of three times on two separate days, this photo must have been taken on either the 8th or the 13th of September 1944. (Photo: *Deutsches Museum*)

Another picture from the flight testing period at Brandis. RATO packs have been installed and the camera ahead of the rudder is pointed toward the areas glued with wool tufts. A few buildings and other aircraft can be made out in the background. (Photo: *Deutsches Museum*)

Taken on the same day, this photo shows the plane shortly before taking off. The machine is being fueled, and hidden behind the front RATO pod can just be seen the starter cart for backing up the internal electrical system needed to turn the energy-hungry engines over.

The Ju 287 V1 seen taking off on a test flight; the RATO pods are still hanging below the aircraft. The film camera ahead of the vertical stabilizer can clearly be made out. Also visible is the absence of the nosewheel spats.

One of the few aerial photographs was not taken, as has been repeatedly claimed, by an Allied reconnaissance aircraft, but by one of the Junkers Ju 188 chase planes climbing to altitude. This photograph shows the unusual flying profile of the Ju 287. (Photo: *Deutsches Museum*)

The Ju 287 V1 seen during landing. With its braking 'chute fully deployed, the heavy aircraft slowly rolls to a stop. (Photo: H.P. Dabrowski)

Another air-to-air silhouette shot taken from an overflying chase plane again shows what an unusual figure the Ju 287 V1 made in the sky. (Photo: H.P. Dabrowski)

Detail photograph of the rudder with the film camera. Another distinguishing feature is the high position of the horizontal stabilizers on the Ju 287 V1.

Up until that point the Allies had been unaware of the existence of this aircraft, and it was temporarily assigned the designation 'Rechlin 66'. After only a brief stay in Rechlin - the entire project had been stopped in the meantime - the Ju 287 V1 was brought back to Brandis without ever having flown at Rechlin. At Brandis its engines were removed and it was parked at the edge of the airfield under some makeshift camouflage netting. Defense of the Reich was now the watchword, and the valuable engines were more urgently needed for the Me 262.

On 8 February 1945 a Spitfire conducted a reconnaissance flight over the area of Brandis. After initial problems the pilot, Flt Sgt K.J.H. Nichols was able to photograph the airfield at Brandis through a break in the clouds. Subsequent evaluation showed that there were two Rechlin 66 types at the airfield. These were the two engine-less Ju 287 testbeds that had been parked there.

Despite the relatively few experimental flights, these were sufficient to confirm the basic usefulness of the negatively swept wing. Other than that, the Ju 287 V1 was in any case only to have been used for basic systems testing.

Yet the Ju 287 V1 can safely boast of wearing a small record on its leaden breast - by logging maximum speeds of 650-680 km/h it is, even today, probably the fastest aircraft with fixed gear ever to fly, and is most certainly the only fixed undercarriage airplane with up to six engines. Furthermore, the Ju 287 was the first heavy aircraft in the world with swept wings and ejection seats.

The second test bed, the Junkers Ju 287 V2 (Ju 288 V202), was to then be used for high speed flight testing, as we shall shortly see.

First Sergeant Gorman D. Gillian of A Battery, 3rd Armored Field Artillery Battalion posed especially for this photo. This picture proves that the Ju 287 V1 was still at Brandis when the Americans took the airfield. The high set elevator bears out that this is the first testbed version. (Photo: M. G. Ruhlen)

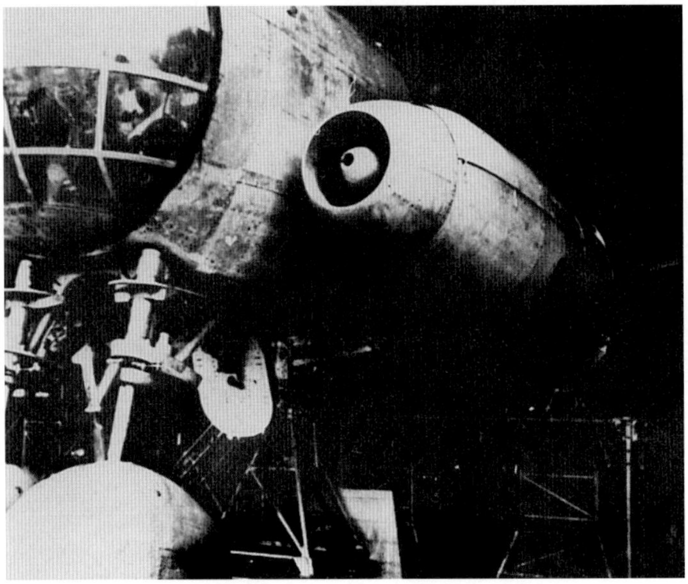

Detail photo of the forward cockpit glazing and the turbojet engine pod nestled against the fuselage.

A front view looking beneath the fuselage offers a good idea of just how massive the main gear was.

Here, too, the photographer 'went underground', but this time focused on the trailing edge of the wing. In this photo the extended landing flap is quite evident.

Another "underground" picture shows details of how the wing engine was mounted.

One last picture from the "underground series" once again shows the nose gear's suspension and the darkly painted "trouser legs" on the Ju 287 V1. The aircraft in the background reveal that the airfield was quite busy at this time.

Junkers Ju 287 V2 Test Bed

Like the Ju 287 V1, the second experimental plane was a composite aircraft constructed from the same components. It differed from the first test bed, however, in minor details: the elevators were set some 30 cm lower, the tailwheel at the end of the fuselage was removed and the opening sealed over, the upper nose wheel undercarriage pants were, unlike the V1, painted in a bright color, and the main undercarriage was braced inward instead of outward as on the V1. Probably the most obvious distinguishing feature were the engines; the Junkers Ju 287 V2 (Ju 288 V202), which carried the registration code of RS+RB, was fitted with a total of six engines. Two Jumo 004 B jet engines were fitted on either side of the fuselage as on the V1, with the other four BMW 003A engines located beneath the leading edge of the wings, two on each wing in a paired unit.

Like the V1, in August 1944 the individual components were shipped to Brandis for final assembly where, as already mentioned, flight testing to a maximum speed of around 800 to 850 km/h was to be carried out. According to *Herr* Walter Rebaum, the Ju 287 V2 was completely assembled to include its engines. Just like the V1 the Ju 287 V2 was rolled out to the calibration hard stand, anchored down, and its engines fired up in order to measure their thrust with a dynamometer. Just a few days later, however, it was rolled with its two fuselage engines and the paired BMW wing mounted engines off to the edge of the airfield, where it was parked and camouflaged - like the V1 - with a makeshift collection of netting and branches. A little while later the engines were removed and transported to Dessau. When it appeared that the flight testing program would be revived, to save testing time it was decided to rebuild the

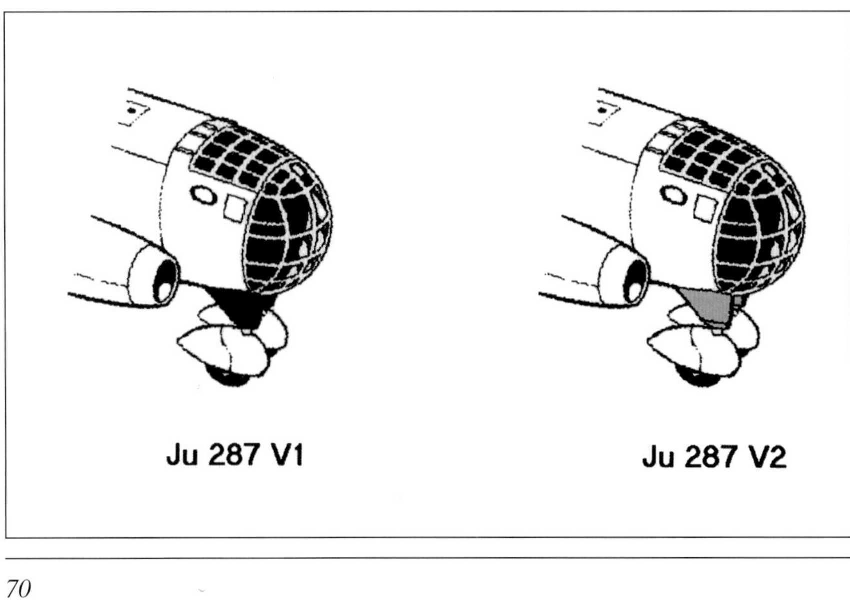

An important differentiating feature of the first two testbeds was the different paint scheme applied to the so-called "trouser legs", or upper landing gear shrouds. On the Ju 287 V1 these were clearly much lighter, probably painted in RLM 02 *Grau*.

This engine-less Ju 287 V2 is seen hastily camouflaged at the edge of the airfield in Brandis. The lack of engines makes the plane tail-heavy, lifting the nose wheels off the ground.

Once again the rather sad looking Ju 287 V2, this time from the front.

Aerial reconnaissance photo taken on 8 February 1945 by Flight Sergeant K. J. H. Nichol over Brandis, clearly showing the "improperly" parked Ju 287 V1 on the far right. Near the center is a "properly" parked Ju 88. The aircraft on the left are wrecks of a handful of He 177s.

engine configuration as a triple pack, as it would be found on the production version with two triple engine combinations under the wings. This intention promised to save time, for then the reconfiguration of the engines could be dispensed with and the testing of the triple engine combination could be tested in conjunction with the high speed trials. There was one slight problem, however: at this point in time a triple engine configuration of the BMW 003A was not available. Since these engines were not to be had, Junkers made a virtue out of necessity and intended to mount a single Junkers Jumo 004 B turbojet under each of the BMW 003 pairs on hand. This has led to some confusion in the research, as all participants who still can recall the events swear blindly that this combination was tested. However, as the photograph on this page shows, it was exactly the other way around: a triple combination was built using paired Junkers 004 Bs with a BMW 003 A slung beneath.

This version is confirmed by one of the former thermodynamicists for jet engines at the Junkers Works in Dessau, *Herr Dipl. Ing.* Theuer, as well as by Walter Rebaum. Both have stated in identical words that an experimental wing of the Ju 287 was constructed at Dessau for the purpose of serving as a test stand for the triple engine combination consisting of paired Junkers engines combined with an underslung BMW 003. But these engine configurations were never fitted to the test aircraft due to the test flights being dropped in late September/early October 1944.

When production was unexpectedly intended to resume in March of 1945, the test flights were dispensed with in any event, primarily due to a shortage of engines.

Another little anecdote that actually isn't one at all perhaps will clear up any possible confusion regarding the two testbeds. The reconnaissance flight mentioned in the previous chapter taken by a British recce plane shows one of the two test aircraft "improperly" parked in the woods adjoining the edge of the airfield. What this means is that as a rule the airplanes were positioned with their tails toward the woods. In one of the photographs, however, a Ju 287 is positioned exactly the other way around (see picture on page 71). The reason for this is quite simple - the ground crew were simply too lazy to take the effort to manhandle the heavy beast around. It is well known that maneuvering the Ju 287 V1 was a very tedious affair. After its last flight it was rolled forward under turbine power into the parking bay at the edge of the woods and was left as it rested. When the engines were needed elsewhere, these were removed and caused the Ju 287 V1 to become tail heavy,

A photo full of symbology. It is good evidence that not everything goes according to plan. Flying in the face of statements to the contrary, the picture irrefutably shows that a triple engine combination consisting of a twin Junkers configuration of two JUMO 004 Bs with a BMW 003 A was in fact tested. The Junkers turbines are easily recognizable by the extended covers for the Riedel starter motor, whereas the BMW turbines had the Riedel starters deep inside the intakes, giving the "mouth" a noticeably darker appearance in comparison.

The unfortunate end of the Ju 287 V2, identified by the lower position of the elevators just to the right of the American G.I. Other characteristics of the Ju 287 V2 include the attachment points for the twin BMW engines still visible here beneath the wing, as well as the inward cant to the landing gear support struts. The entire forward fuselage has been blown up and hangs in tatters. The G.I. mentioned earlier proves that both testbeds were in Brandis at the time the Americans arrived.

The partially destroyed Ju 287 V2 taken from the forest toward the airfield.

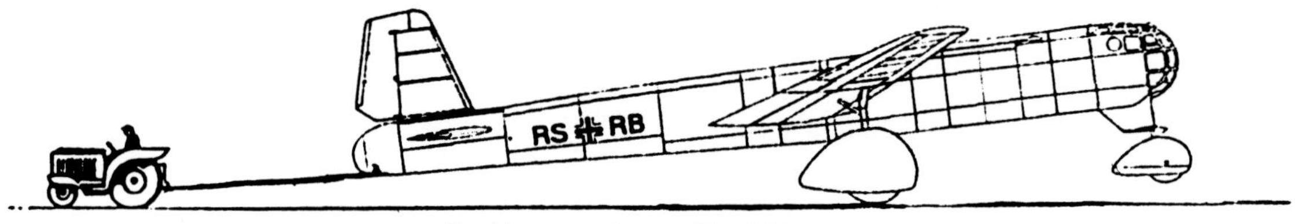

This sketch shows the method for towing the engine-less Ju 287 V2 from the tail. With no engines, the tail-heavy machine could be towed in this fashion without difficulty.

tipping back onto its tail skid. So far, so good - but then why were the V1 and V2 both positioned correctly at the war's end? Well, the Ju 287 V2 had also had its engines removed, causing it to be tail heavy like the V1. For this reason the test aircraft were towed by their tails using a tractor, which could maneuver the behemoths quite easily now that their nose wheels were no longer on the ground (see picture below). The V2 was therefore towed to the edge of the airfield tail-first and parked "properly" there. The little tractor was easily able to drive beneath the wings. Sometime during the month of March 1945, when the program started up again, the V1 was towed to the large assembly hangar in the same manner, with the expectation that testing would soon begin again. In this event, the V1 would be the first to be certified flight ready. Yet when it became apparent that these hopes were premature and the test flights would not occur for either the V1 or the V2, the V1 was transported in the usual way to the airfield's edge and parked, as was the standard for the *Luftwaffe*, with the tail toward the trees. As you can see, not really worth mentioning, but interesting nonetheless.

The reason that any renewed testing was out of the question most certainly came down to the fact that it was assumed the Ju 287 V3 would soon be available for trial flights as the prototype, and prototype testing could be undertaken directly. This would obviate the need for any further experiments with the two testbeds - they simply weren't needed anymore.

Thus, as the war drew to a close the two testbeds stood, stripped of their engines, at the edge of the Brandis airfield. As the Americans approached, several explosive charges were placed in the cockpit areas of the two testbeds and, with heavy hearts, the charges were detonated. Nevertheless, the Americans were able to capture the two torsos of these test aircraft with their wings, the bulk of the fuselages, and the empennages intact.

As of now, it is not known whether the Americans or later the Russians took these parts, or whether the remains were possibly used for static testing, e.g. for the project studies for the EF 131, or whether the parts were simply scrapped. Perhaps a reader might know more.

Production Prototypes, Junkers Ju 287 V3 and Other Prototype Versions

When the whole jet bomber program was halted in the autumn of 1944 the Ju 287 V3 was already in an advanced stage of construction. Often, though, during the last months of the war higher authority would give short notice instruction directing that canceled projects be restarted, and it was because of this that work on the program did not cease immediately, but continued doggedly apace. Once it became apparent, however, that the stop in construction was not temporary, all activity for continuing the project eventually ground to a halt.

As late as 9 September 1944 Junkers compiled a status report on the plane's development:

> The layout of the Ju 287 has been based on a normal bomb load of 3,000 kg, with a range of approximately 2,000 km for non-rocket assisted takeoff and with an overload of 4,000 kg a range of 1,900 km. Assisted takeoff, consisting of three WASAG powder rockets of 1,000 kg thrust with a 12 second burn time, is also needed when engine thrust is reduced due to high temperatures.
>
> With these assumptions, the design has an all-up weight of approximately 22,800 kg with an FHL 131/Z tail barbette based on a wing surface area of 58.4 m^2. In order to get the performance with the powerplants currently available, it is planned to install six BMW 109-003 A1 engines. Installing six Jumo 109-004 B1 or C engines is possible without the need for any serious modifications to the airframe.
>
> The unique features of the turbojet engines necessitate the aircraft flying at the highest possible altitudes to ensure adequate economy. For this reason installation of dive brakes and a dive bombing sighting system have been dispensed with for the time being. Armament consists of an FHL 131/Z tail barbette, which is operated by the radioman via a PVE 11 periscope sight coupled with an FA 15 remote drive unit. The crew is protected from the rear by sensibly placed armor plating. Approximately 80% of the fuel is carried in 4 mm thick bladder tanks that offer some protection against attack.
>
> With regard to strength, a load factor of nsi = 4g/-1.5g has been calculated at a maneuvering weight of 18,500 kg. The load factor for the engine is nsi = 5g and for the empennage nsi = 4.5g.
>
> A future follow-on development of the aircraft is possible by taking the given sweep angle and swapping out the outer wing sections for larger chord ones with new flaps and ailerons, and the wing tips for an extended span tip, giving the potential for a wing surface area increased to approximately 68 m^2. The wing center and leading edge sections would remain unchanged.

Primary Aircraft Data

Wing area	58.4 m^2.
Wing span	19.4 m
Wing sweep @ 50% of the wing	-23.5%
Wing dihedral	8.5%
Takeoff weight	22,780 kg
Powerplant	6 x BMW 109-003 A1 in two triple mounts
Crew	Three
Armament	Tail: FHL 131/Z with 2 x 600 rounds
Armor	Crew protection from behind
Sighting System	Lotfe C 7 H (later Lotfe and PVE 11)

Junkers Ju 287 V3:

But let us return to the Ju 287 V3. The third prototype was thus given an entirely new airframe. Planned engines were the two BMW triple combinations, each consisting of three BMW 109-003 A1 engines. In March 1945 new life was breathed into the jet bomber development program, so that the work that had earlier been frozen could, after a short preparation time, be renewed with vigor. Nevertheless, despite combining the developmental and construction efforts into a single priority action, the Ju 287 V3 could not be completed before the end of the war. Thus, when the Americans overran Dessau, they found the Ju 287 V3 to be about 80-90% complete. After handing it over to the Russians, the new masters had all the engineers, technicians and other experts who had participated in the Ju 287 program brought together. A new construction group was formed, which initially was tasked with finishing the Ju 287 to the point that it could be used as a stress testing airframe for additional static experiments. Not such an easy undertaking, as a lot of material had been destroyed at the last minute. Nevertheless, all the components in their various stages of completion they could get their hands on were brought together. Of course, since the V3 was not expected to fly, it wasn't necessary to finish installation of any functioning instrumentation, for example. The next step was for the group to reconstruct the Ju 287 based on the remaining fragmentary documentation. This eventually spawned the EF 131, but that is a story in itself and will be explored in more detail later.

Any idea that the Ju 287 V3 flew in Russia after the war is nothing more than a fairy tale, and not even the fantastic air brushed pictures of Gert W. Heumann can change this. These show the V3 flying through the air, sometimes with German markings, sometimes with Russian. Furthermore, the V3 would never have flown with armament and the engine configuration shown. But, in defense of Gert W. Heumann, it was he who attempted to make the

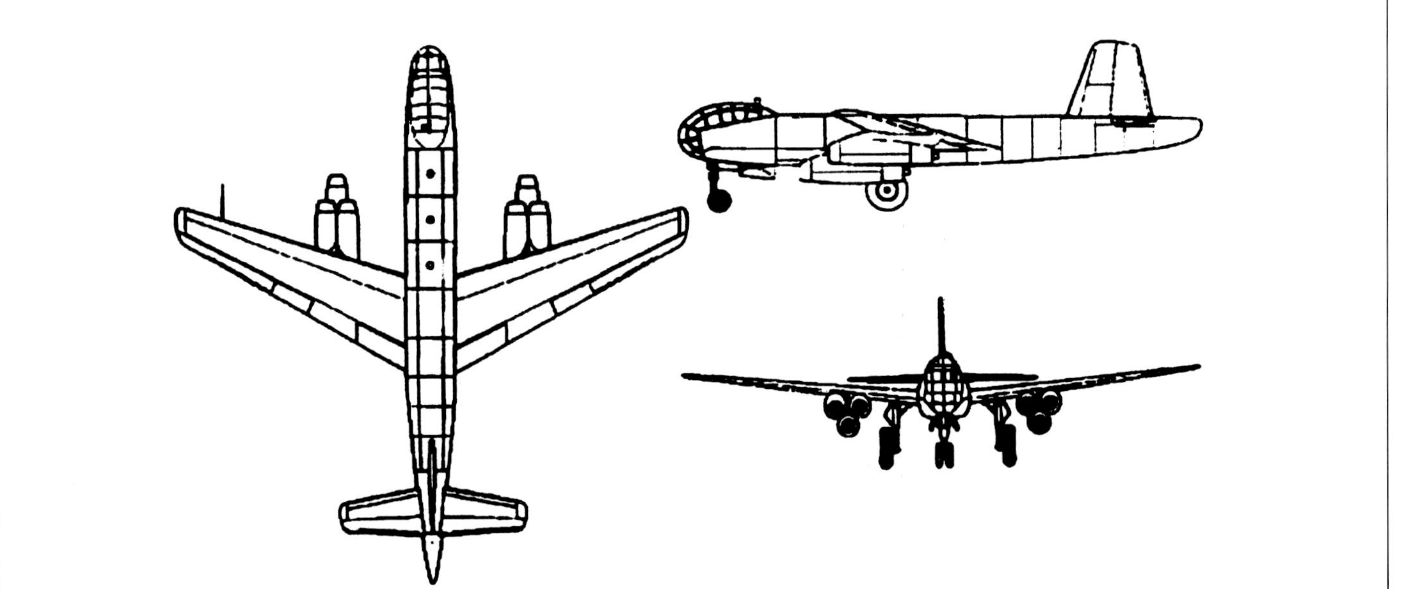

An early three-view of the Junkers Ju 287 V3 minus the remote gun turret in the tail.

Another of Heumann's famous pictures. This 'photo' can be found in virtually every publication. Even today, this picture is repeatedly foisted upon the believing reader as a 'real' photograph. There are several variations to Heumann's picture, sometimes in German markings, sometimes in Russian (as evidence that the Ju 287 V3 flew in Russia). In spite of this, neither the Ju 287 V2 - which the picture is sometimes claimed to represent - nor the Ju 287 V3 ever flew with such an engine configuration, not to mention the fact that they never got off the ground in the first place.

developments of the day palatable to a broad readership base in an era when virtually nothing was known about the so- called secret weapons. His pictures, many of which can barely be discerned from the real thing, are truly unique - and this in an age when there were no computers. Nevertheless, some of these pictures contained a small yet obvious error: they did not always correspond to reality - a fact that many authors simply ignored. Therefore, today it is not so easy to correct information that has been ingrained over decades and get it out of people's minds. Once again, the Junkers Ju 287 V3 *never* flew.

Mockup of the crew compartment of the Junkers Ju 287 V3 as built by Junkers.

The spacious cockpit for the Ju 287 V3's crew seen from the front quarter shows once again the generous glazing and the resulting excellent all-round visibility it provided.

Far Left: View from above into the mockup of the Ju 287 V3 cockpit already showing planning for several instruments, the ejection seats, etc. Left: Front view of the Ju 287 V3 cockpit.

This photo shows the aft section of the crew compartment with the mockup for the periscope sight.

Triple engine combo on the test bench attached to the evaluation wing of the Ju 287 V3.

Junkers Ju 287 V4:

The Ju 287 V4's layout was the same as the V3's with regard to airframe and engine and was also to have served as the prototype for the production version. The V4 was also under construction at the war's end and had been about 60% completed.

Junkers Ju 287 V5:

The fifth prototype also had the same airframe and engine configuration as the V3. But it was the first machine to make use of armament in the form of an automatic FHL 131/Z tail turret. Furthermore, it was the first prototype aircraft that had an ejection seat system fitted in the cockpit. Some of the components had been completed before the end of the war, such as the FHL 131/Z automatic barbette.

Junkers Ju 287 V6:

The V6 was most likely the last prototype whose construction began during the war; in every way it was virtually identical to the V5.

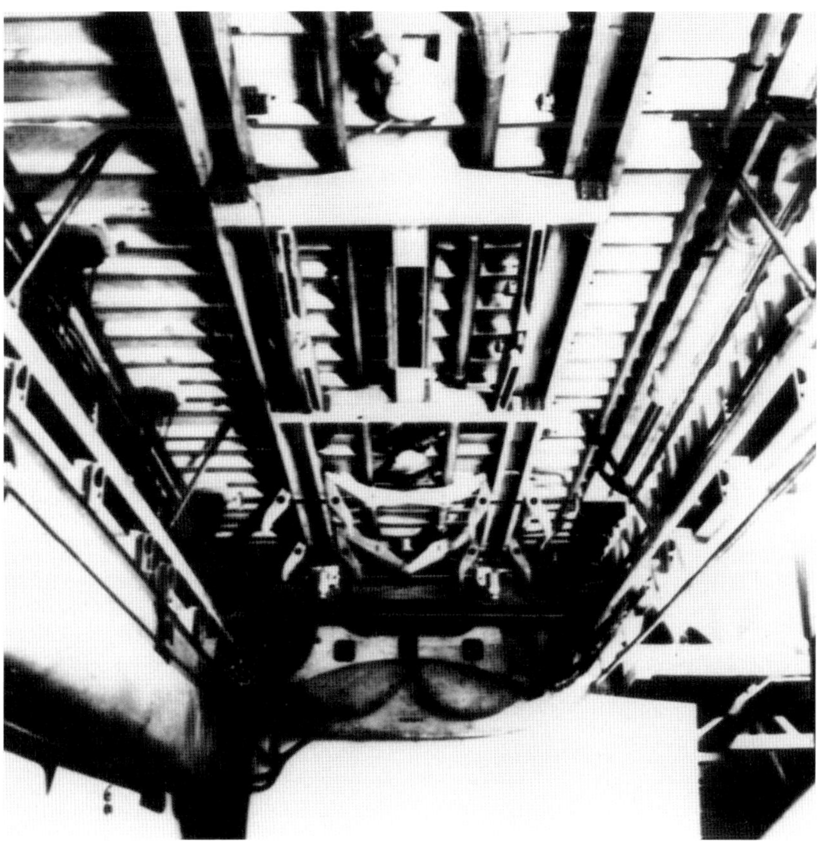

View into the Ju 287 V3's open bomb bay.

Pre-Production:

Up to 154 pre-production aircraft were to supposedly to have been built. Not only did production on these never begin, but pre-production numbers were never even allocated. These aircraft were identical to the V5 in terms of equipment, but were initially not to have been fitted with the ejection seats.

Junkers Ju 287 A1 Series:

In addition to the prototypes and the pre-production versions there were two other main series that would have been built. The machines of the A1 series again would have corresponded to the V5 in terms of configuration and equipment, but also would not have used the ejection seats.

Junkers Ju 287 B1 Series:

The B1 series differed from the A1 series only in a few details, such as two better performing triple combinations replacing the BMW triple engines. These would have comprised three Jumo 109-004 C engines. Other differences included larger 1550 x 575 main undercarriage wheels as well as a larger 1,015 x 380 nose wheel, with the

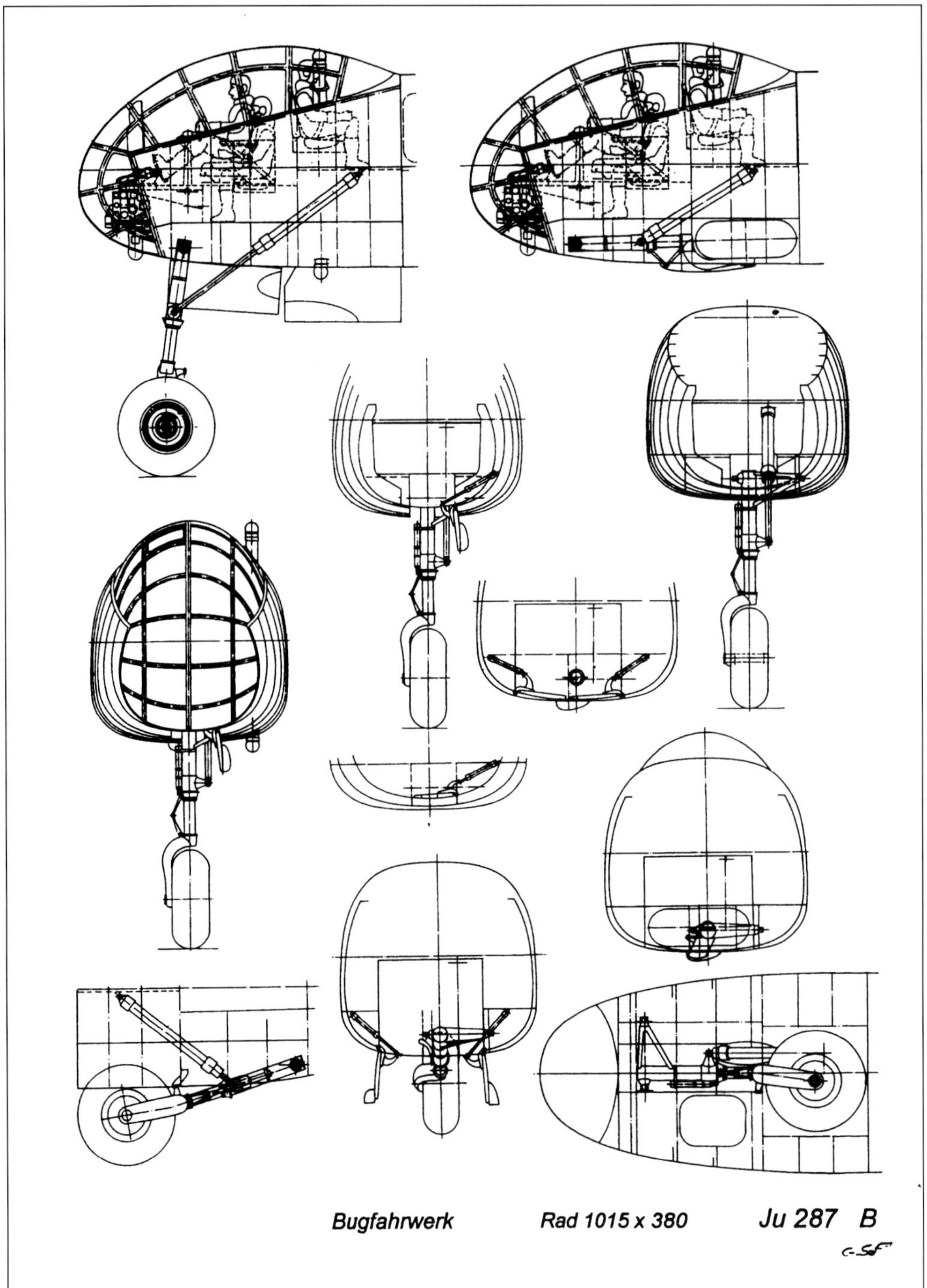

Design drawing for the nose gear of the planned B version with larger tires and a single wheel. (Drawing by Günter Sengfelder)

takeoff weight increasing to 25,200 kg. Otherwise the Ju 287 B1 was the same as the Ju 287 A1.

Other Testing:

In order to save time and expense, individual components were often tested on existing aircraft types.

According to Junkers documents, a Ju 188 G2 was accordingly converted to test the original tail configuration of the Ju 287 and was fitted with the empennage planned on production models. By so doing, it became possible to apply the data from the experimental flights directly to the production line without the necessity of flying a prototype or waiting until one was available.

At this point it is worth taking a closer look at the Ju 188 G2. It has been repeatedly claimed that the Ju 188 G2 never existed, but this is only partially true. The Ju 188 G2 was already in production when it was decided to abandon production in favor of the Ju 388. A large part of the production was subsequently scrapped. But about two or three machines that were almost built were probably unofficially completed and used as company testbeds. Components from the Ju 188 G2 production were also used for the two Ju 287 V1 and

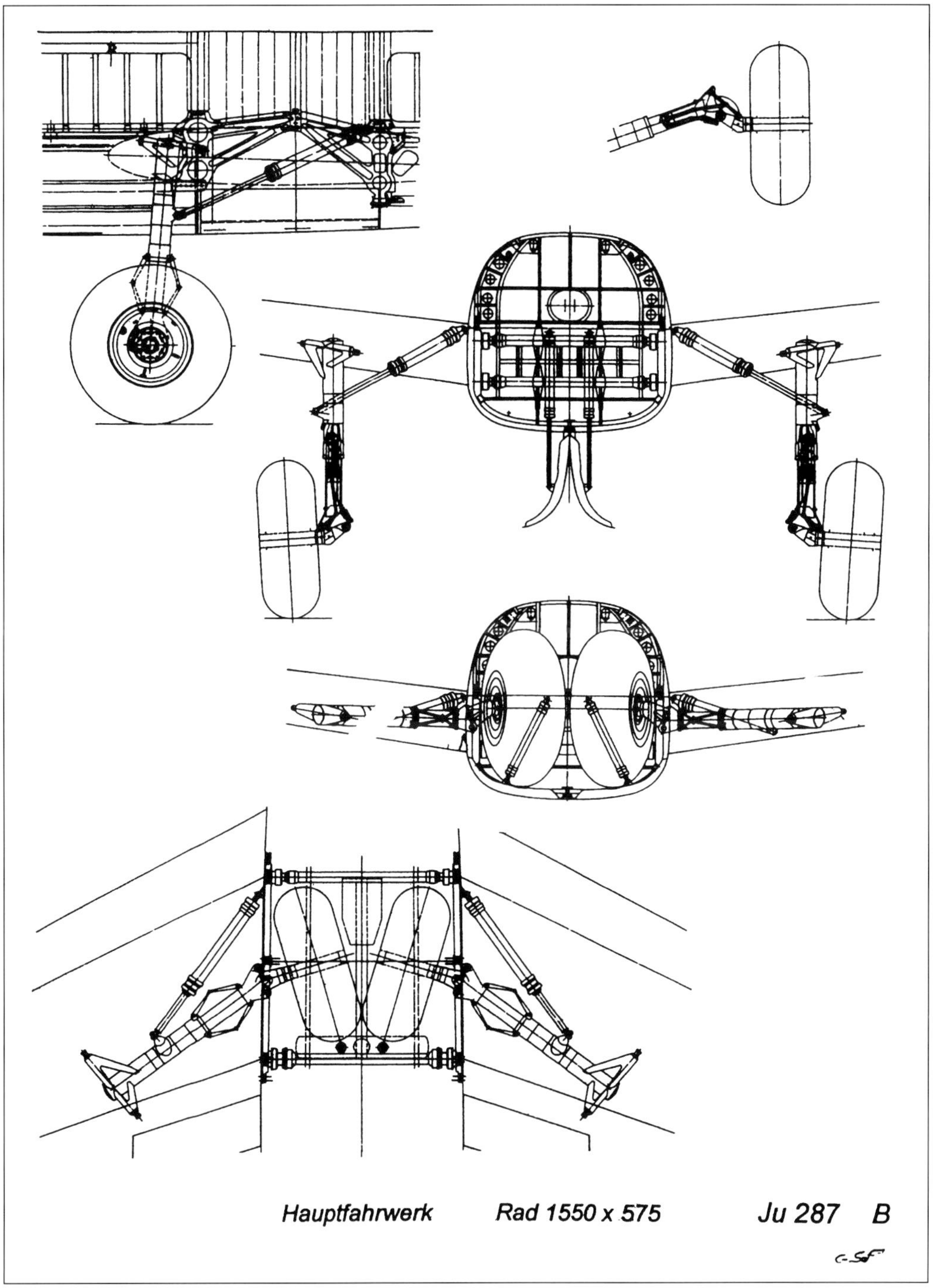

Design drawing for the main undercarriage for the planned B version of the Junkers Ju 287. (Drawing by Günter Sengfelder)

Test framework for the Ju 287's nose gear system, here fitted with skis for trials in the snow in the early part of 1945.

V2 testbeds. These Ju 188 G2 parts and testbeds are documented in Junkers records and the photo on the next page below most likely shows the empennage of a Ju 188 G2.

Additional experiments were conducted using the Heinkel He 177 V38, coded KM+TB, whose bomb bay was enlarged accordingly to evaluate the bomb release system of the Ju 287. After the war's end, this test bed lay on the scrap pile and a few misguided authors of questionable aviation literature saw in this wreck the first German atom bomber (see photos below and on page 83).

Another Heinkel He 177 was equipped with the original undercarriage of the Ju 287 for testing this component. Furthermore, for evaluating the nose gear retraction hydraulics, a special test sled was constructed in order to enable testing the twin-wheeled nose gear.

To be sure, all these experiments provided important information, but could not be fully exploited due to the rapid end of the war.

Remains of Heinkel He 177 V38, *Werknummer* 55000002, at Prague- Ruszin. This served as the testbed for the overlarge bomb bay of the Ju 287. Oftentimes passed off in many publications as the German "atomic bomber".

A look into the oversized bomb bay of the He 177 V38, the reputed "atomic bomber". These claims most probably are derived from the fact that the airplane was in Prague at the war's end, and that Prague at the time was home to all kinds of "wonder weapons". Certainly no one could understand any reason for an He 177 to have such a large bomb bay other than it must have been an "atomic bomber".

Another photo - this time from a different angle - of the brutally plucked wreck of the He 177 V38, the testbed for the Ju 287's bomb bay.

In the picture's foreground is a Ju 88 S1. In and of itself nothing particularly special. On the left side, however, can be seen a rudder that isn't quite as easily classifiable. It bears a remarkable similarity to the rudders of the two test platforms. Perhaps a Ju 188 G2???? Can anyone provide more details?

First Operational Plans for the Junkers Ju 287

Plans were being drawn up for Ju 287 operations even during the developmental phase, when it hadn't been determined whether and in what numbers the Junkers Ju 287 would be built. Taking into account the existing operational bomber types, on 8 July 1944 a new program was drawn up outlining the distribution of the different aircraft types among the individual bomber units.

What is noticeable is the fact that the future bomber groups would be entirely stocked with the new propulsion technology in the form of jet engines; only the Dornier Do 335 appears in the overview as the single propeller driven type. The other propeller driven machine of any significant numbers, the Junkers Ju 388, is a transitional proposal. In this overview plan for rebuilding a bomber

The modified model kit box cover from the Planet Modellbau company reputedly shows an operational flight by a Ju 287. Like the often mentioned "Heumann picture" of the Ju 287 that keeps haunting us, this too shows an incorrect engine configuration.

air force (based on delivery plan 226/2), dated 8 July 1944, there were already two bomber groups planned for having the Junkers Ju 287, each with 37 aircraft.

Those bomber groups that still operated propeller driven aircraft such as the Heinkel He 177, the Dornier Do 217, the Focke Wulf Fw 200 etc. were to be disbanded once their existing aircraft inventory had been 'used up'; for the most part, only the so called 'jets' would see service in the future.

This overly optimistic prognosis operated on the assumption that establishment of the two Ju 287 bomber groups could, amazingly enough, begin in May of 1945 and be completed by December of that year.

As a classic strategic bomber, the Ju 287 was to have primarily operated from higher altitudes and dropped its large bomb load from level flight; only for certain special operations would a high speed gliding attack profile have been considered. It was assumed that the Ju 287 would enjoy a long term advantage over Allied fighters in terms of speed, enabling it to effortlessly outpace them. To protect itself from the rear on takeoff and during landing, a remote controlled tail gun barbette sighted via periscope was considered entirely adequate. For the time being, additional gun turrets typical for a bomber's defenses were felt to be unnecessary due to its tremendous speed advantage.

The wide variety of bomb types and calibers capable of being carried made the Ju 287 a true universal bomber. The different loadouts can be found on the overview table on pages 86 and 87. These ambitious plans could not be realized due to the war, and the opportunity for making any technological advancement was again lost by the halt in construction. On the other hand, it is highly doubtful Germany would have been able to afford such an expensive project in view of the increasingly grave shortage of materials. Yet again, here was another example of a program being thwarted by a senior leadership completely out of touch with reality, whose decisions reflected the inability to differentiate between the feasible and wishful thinking.

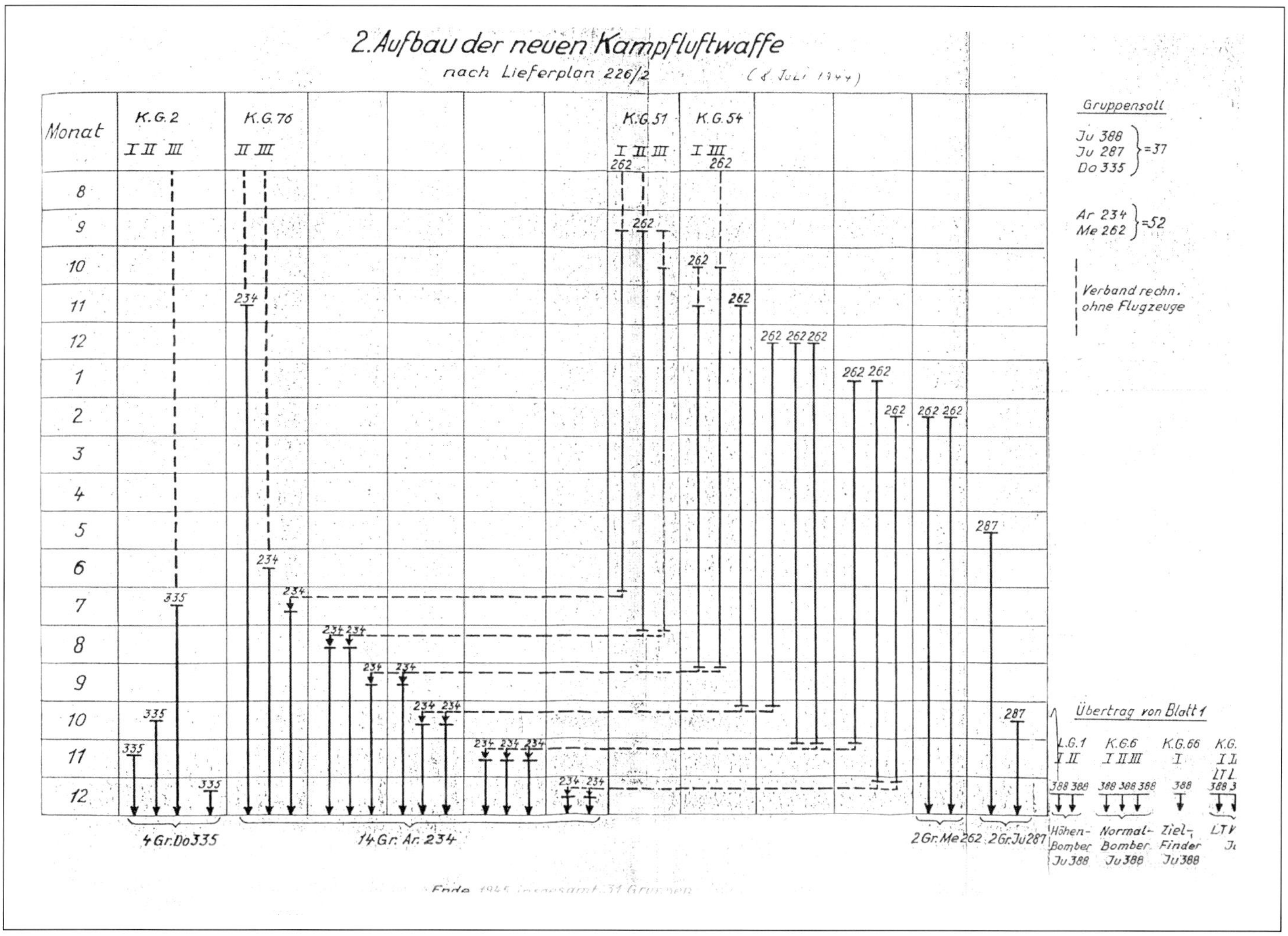

Diagram of the newly organized "Bomber *Luftwaffe*" through the end of 1945 showing a total of 31 *Gruppen*. Notice that very few propeller-driven aircraft would continue in use - the overwhelming majority of intended aircraft would be jets.

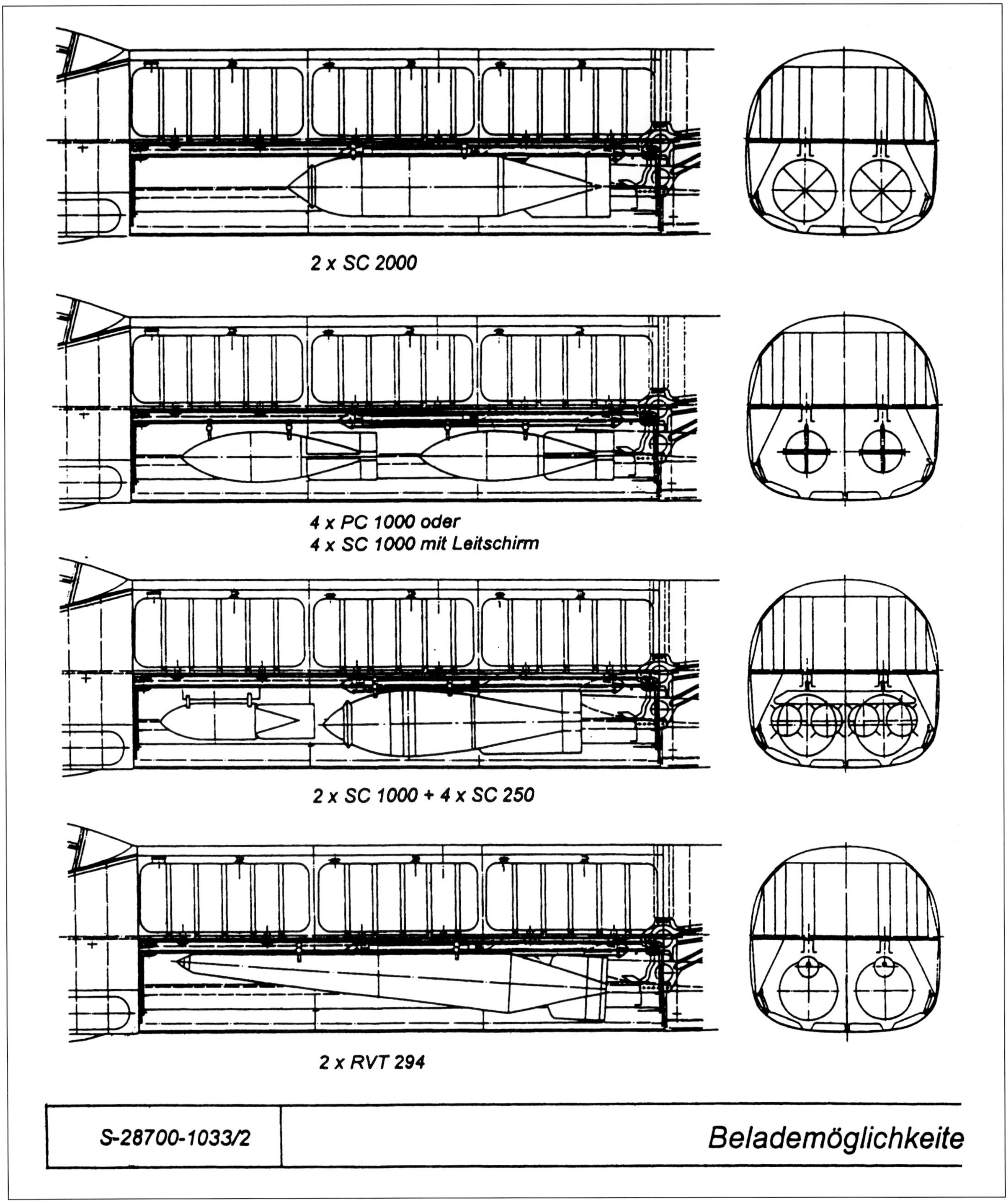

Potential load-outs for the bomb bay of the Ju 287. (Drawing by Günter Sengfelder)

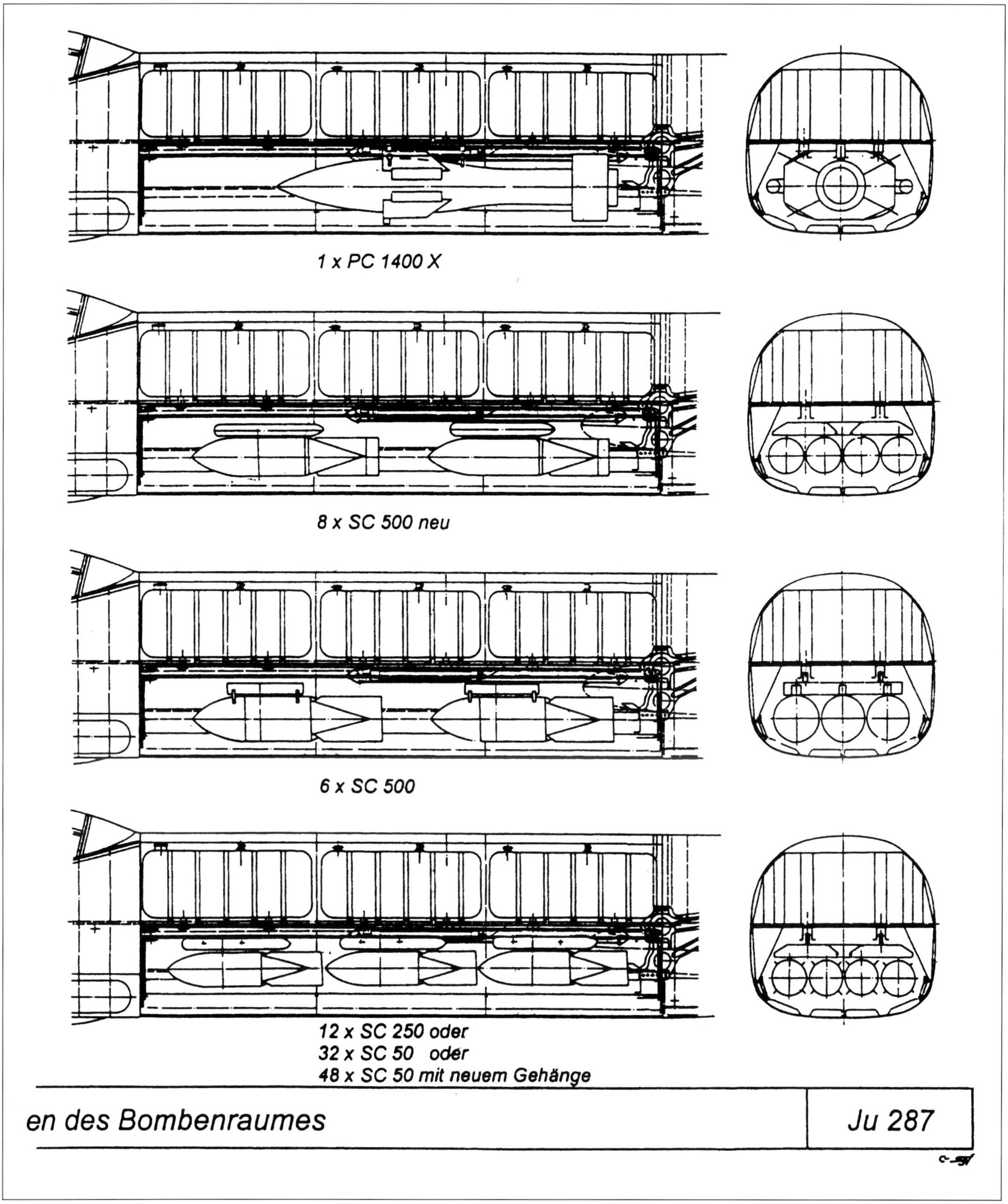
1 x PC 1400 X
8 x SC 500 neu
6 x SC 500
12 x SC 250 oder
32 x SC 50 oder
48 x SC 50 mit neuem Gehänge
en des Bombenraumes
Ju 287

The model in 1/15th scale by Günter Sengfelder shows a Ju 287 prior to a fictional combat mission. The Ju 287 is just being loaded with a remote-controlled SD 1400 "Fritz X" free-fall bomb.

Another photo of the model, virtually indiscernible from the real thing, showing a Ju 287 at the runway's edge in preparation for takeoff. Notice the WASAG 109-522 RATO packs fitted to the underside of the aft fuselage.

Production of the Junkers Ju 287 "*Gewaltaktion 287*"

After our short intermission with operational planning, we now turn our attention again to a more concrete topic: production of the Ju 287.

Even during the Ju 287's early development phase plans were being hammered out for mass production of the type. The initial production batch was to have consisted of 100 units of the first version. It was obvious to everyone involved that such a highly complex, high-tech operation (as it would be called today) necessitated the involvement of other facilities besides just the parent plant in Dessau, and the need for underground production/fortified structures was recognized early on. At this point in time, the Junkers Works was chiefly organized into the following concerns (*see* below).

For going underground, prospects included nearby subterranean factories such as the Thyrawerke facilities near Rottleberode as well as the so called Nordwerk near Niedersachswerfen. This was the first step in protecting at least a portion of the Junkers Works against bombs.

But Hitler, too, had his own ideas about bomb-proof aircraft factories. As we know, Hitler always went into raptures whenever enormous amounts of concrete were used. So it was no wonder that he proposed the construction of two massive aboveground aircraft production facilities that, according to his plans, would consist of two gigantic concrete blocks, each with a usable floor space of at least 600,000 square meters. Labor for these fantastic constructions would be provided by 100,000 Hungarian Jews. Hitler

Aircraft production:	parent plant:	Dessau	**Engine production:**	parent plant:	Dessau
	branch:	Aschersleben		branch:	Kassel
	branch:	Bernburg		branch:	Köthen
	branch:	Halberstadt		branch:	Leipzig
	branch:	Leopoldshall		branch:	Strasbourg
	branch:	Schönebeck		Otto Mader Works:	Dessau
	Langenwerke AG:	Langensalza		Zitterwerke AG:	Zittau
	construction yard:	Breslau			
	construction yard:	Leipzig			

rumbled when his ideas for these giant bunkers were explained to him as being unfeasible during the course of a meeting on 6 April 1944. It took considerable effort to persuade him that the more practical and more economical solution would be to house the production facilities underground by expanding the tunnel network of the Mittelwerk plant in Kohnstein, then under construction.

When development then stopped in September 1944, the ambitious production plans for the Ju 287 were shelved for the time being. The situation changed, however, a few weeks following the assassination attempt on Hitler on 20 July 1944. Hitler no longer trusted even his closest circle of advisors within the *Wehrmacht* and now relied more heavily on the support of SS. *Generaloberst* Fromm, Chief of Army Weapons and Commander of the Reserves, was arrested as one of the conspirators and removed from his post. SS *Obergruppenführer* Max Jüttner was appointed in his place. Thus, the SS became increasingly influential with regard to the fate of aviation development, among other things. Within the ranks of the SS there were dreams of having their own *Flieger-SS*, or aviation branch. The *Luftwaffe* was considered inadequate and completely outdated. It was intended to equip these newly organized forces with the most modern equipment, effectively training them as high-tech fighters. Physically of good constitution, fearless, daring and victorious in combat - this was the ideal vision of these SS airmen. On 7 June 1944, for example, SS *Standartenführer* Dr. med. Martin Brustmann wrote to SS *Hauptsturmführer* Dr. May:

> ...if there is ever to be our own SS aviation, then the economical engine we're talking about will become burdensome. We have to have our own SS aviation, since with the *Luftwaffe* we're stuck with such outmoded ideas that we can't really expect any fundamentally new developments from that side that would guarantee for us another decisive advantage.

For the members of the *Flieger-SS*, only the most modern aviation equipment would be provided. Another reason why the Ju 287 - after testing was, without any ado, suddenly halted in the fall of 1944 - again found itself at the top of the SS's most wanted list was due to the comet-like rise of SS *Obergruppenführer* and General of the *Waffen SS, Dr. Ing.* Hans Kammler. In order to cast a bit more light on the reasons behind the sudden resurrection of the jet bomber, it is necessary to take a somewhat closer look at the person of Hans Kammler and the impact of the SS. At a relatively early stage, the SS leadership realized that to attain unlimited power, more than just fear and terror through brutal and ruthless measures was needed. "Intelligence", said the scientists, was also a vital ingredient. There were dreams of an all-powerful SS think tank that would develop fantastic, never-before-seen weapons designed to force any potential enemy to its knees in short order. An enthusiastic champion of this idea was *Dr. Ing.* Hans Kammler who suddenly materialized from thin air to advance in a remarkably short period of time to what was virtually the second most powerful man in the Third Reich. Hitler himself empowered him to this in a general letter sent on 3/27/1945. The contents of the letter are as follows:

The use of jet aircraft has proven to be absolutely superior to enemy aircraft. It now depends on exploiting this advantage as quickly as possible by readying quantities of jet aircraft sufficient to break the terror from the air. This can only occur by combining all available resources of the Reich under a strict leadership. This task has priority over all other military and economic matters. To reach this goal, I have tasked SS *Obergruppenführer* and General of the *Waffen SS, Dr. Ing.* Hans Kammler with the following:

1. Command of all jet aircraft development, testing and production to the point of operational service and those materials from the Reich Minister for Armament and War Production needed for operations.

2. Command of all prerequisites in the area of the Reich Minister for Aviation and Commander in Chief of the *Luftwaffe* needed for jet aircraft to the point of operational service.

3. The authority formerly in the hands of the Reich Minister for Armament and War Production relating to jet aircraft now rests with SS *Obergruppenführer* and General of the *Waffen SS, Dr. Ing.* Hans Kammler. Effective immediately, the Commander in Chief of the *Luftwaffe's* designated authority for jet aircraft is now subordinated to SS *Obergruppenführer* and General of the *Waffen SS, Dr. Ing.* Hans Kammler.

4. SS *Obergruppenführer* and General of the *Waffen SS, Dr. Ing.* Hans Kammler is personally subordinated to me for the execution of this tasking and has full authority thereto. He has at his disposal all command centers, departments and facilities of the *Wehrmacht*, the Party and the Reich, who are to follow his directives.

The Führer
(signed) Adolf Hitler

With this letter Hans Kammler was factually empowered to use and employ according to his will and whims all facilities in the Third Reich at the time. Hitler had lost faith with all other of his most trusted advisors up until then. This letter not only deprived of power Armaments Minister Albert Speer and *Reichsluftmarschall* Hermann Göring, but also practically all important positions of every kind in the Reich. But Hitler's switch with regard to the importance of jet aircraft came to late to have any effect. For a long time, Germany no longer had been capable of changing the course of its fatal and hopeless military situation. As early as the fall of 1944 the SS leadership had pushed for taking over control of development of secret and special weapons. And so the Ju 287 inevitably came under the influence of the SS.

Who was this Hans Kammler, what set him apart that he was able to go from being an almost insignificant engineer doctorate and aircraft builder to the highest ranks of trusted associates of Himmler and even Hitler in such a short time? He first drew attention to himself by completing a number of projects he'd been entrusted with in impossibly short periods of time. Without any consideration for the value of human life he employed draconian measures to implement his projects. A combination of brilliant technological knowledge and a talent for organization made him the Nazi leadership's ideal accomplice, and he was soon promoted to commander of rocket and jet development. His ideas and the implementation of them were inimitable and left even the armaments minister, Albert Speer, standing in the shadows. It was he, for example, who drew up the idea for building the underground aircraft and rocket factory , the Mittelwerk near Nordhausen, and Operation Olga, the secret expansion of the "Jonas Valley" also bore his mark, as did the previously mentioned "Military Technological Think Tank" within the Skoda Works in Pilsen. The most wonderful weapons have been attributed to this think tank (advanced research in the fields of atomic weapons, laser weapons, jet aircraft, rocket weapons, flying saucers, etc.), although the truth of these claims must be treated with caution, but that is another story altogether.

Hans Kammler was one of the most secretive figures in the Third Reich; his trail was lost in the murkiness of the war's final days. In his book *"Teufel oder Technokrat"* (Devil or Technocrat), Tom Agoston even went so far as to call Hans Kammler Hitler's gray eminence.

Alongside other projects, such as the V-weapons, Special Unit Kammler worked on a plan that called for gigantic production facilities for the sorely battered aviation industry. They would house jet fighters, jet bombers, jet engine construction and more of the like. For the Junkers Ju 287, a tunnel system using the covername of "Kaolin" was envisioned. Called Project B12, it would be located near Woffleben and the details were worked out under the direction of the Prof. Dr. Rimple architectural firm. *SS-Obersturmführer* Schlachter was assigned overall responsibility for the project. This Project B12 was to have been one of the 600,000 square foot factories that Hitler had called for. *SS-Obersturmführer* Schlachter arranged for a work camp to be established on 3 January under the codename "Erich" so that work could begin on Project Kaolin. Personnel numbers rose from 242 work prisoners to about 1,600 by the time it was disbanded on 4 April 1945. Working conditions were inhumane, yet again and again a few managed to escape from this nightmare. By the time the war ended, only about

One of the few existing photographs of Hans Kammler.

10% of the necessary space had been chipped out of the rock. Thus, the underground facilities for the Project B12 were never used for the planned production of the Ju 287.

Once Kammler had the entire V-weapons development program in his hands, he began working on acquiring suitable equipment for his *Flieger-SS*, then being built up. In addition to the increasingly necessary defensive weapons, such as the manned Bachem Ba 349 anti-aircraft rocket (see bibliography page 219, No. 61 and 62), he vehemently supported the resumption of work on the Ju 287. It was expected that production maturity would be attained through a priority action program. Such a program had previously been implemented successfully for the Heinkel He 162, the so called *Volksjäger*, or People's Fighter, as well as to some extent for the development and full-scale production of the Bachem Ba 349 *Natter*. For those involved, this meant longer work times, greater pressure to succeed in the shortest amount of time without regard for human needs, coarse approaches to acquiring necessary materials, etc.

As described earlier, production was to have been undertaken in the caverns of the Mittelwerk near Woffleben, but the war's events precluded these plans. A last desperate attempt to initiate production was nevertheless begun. In order to cheat fate at least somewhat, production of individual parts was distributed among several smaller operations with a view to avoiding the ever present danger of bombing. These companies had no idea for what ultimate purpose they were building the parts. These parts were to then be put together into subassemblies. The subassemblies in turn would then be assembled in two or four to-be-determined central factories. Contracts had been awarded and production of individual components had gotten well underway shortly before the war ended, and it was hoped that the first Ju 287 bombers could be delivered by mid 1945. In April 1945 American troops occupied the Junkers Works in Dessau, which officially ended production of the Ju 287 without a single operational machine ever having been completed.

As was briefly mentioned earlier, the initiator of the Ju 287 priority action departed for Pilsen shortly before the war's end. His trail was lost there. So, just as he seemingly appeared from nowhere, Hans Kammler disappeared into the mists of the postwar confusion without a trace. Countless rumors have been woven around his reputed demise, but none of these have been able to be confirmed through solid evidence. There is much to support the idea that he, like many of the Nazi bigwigs at the war's end, through Vatican financial aid, was able to make his way to South America or even the United States, where either unrecognized or with the knowledge of the US he continued working on rocket projects. But back to the Ju 287. Disregarding all the positive or negative accompanying symptoms during the Ju 287's development, everything was in place for its production to begin. In order to provide the contractor, the RLM, with a taste of what had been developed and what performance could be expected, a "Brief Description of the Ju 287 Jet Bomber" was drawn up on 19 September 1944, which, curiously, was just before the official termination of work on the project:

> Aerodynamically, the creation of the Ju 287 was approached with a regard for higher in-flight speeds. This resulted in the requirement for thinner wing profiles with delayed onset of the maximum chord and sweep of the wing. Negative wing sweep was chosen because this offered a simple way to improve stall handling on an aircraft with sharply tapered wings on the one hand, and on the other hand it provided a sensible way to make use of the fuselage space. The engine configuration was based on measurements taken in standard and high-speed wind tunnels and was configured so

that a triple engine configuration was mounted on the wing leading edge of both the port and starboard wings. From a design standpoint, the requirements for a high speed aircraft have been taken into account chiefly through the optimal use of space in the most drag efficient bodies possible. A body of rotation of a predetermined slenderness ratio offers the most favorable interior spacing with regard to drag and Mach number. This means that for the fuselage design, the fuel, bombs and equipment can be fitted with minimal dead space. The bomb bay has been designed to accommodate a bomb load of 4,000 kg, since the high fuel consumption of the jet engines (on average 3.5 kg/km) permits a significant increase in bomb capacity with only a minor reduction in range.

Potential bomb load capacity:

2 x SC 2000	4,000 kg
4 x SC 1000 with parachute retardation	4,000 kg
4 x PC 1000	4,000 kg
8 x SD 500 with PARACHUTE RETARDATION	4,000 kg
8 x SC 500 (new type)	4,000 kg
6 x SC 500 (old type)	3,000 kg
2 x SC 1000 + 4 x SC 250	3,000 kg
2 x SC 250	3,000 kg
2 x LT-RTV 294	3,000 kg
48 x SC 50 with new rack frame	3,360 kg
32 x SC 50	2,250 kg

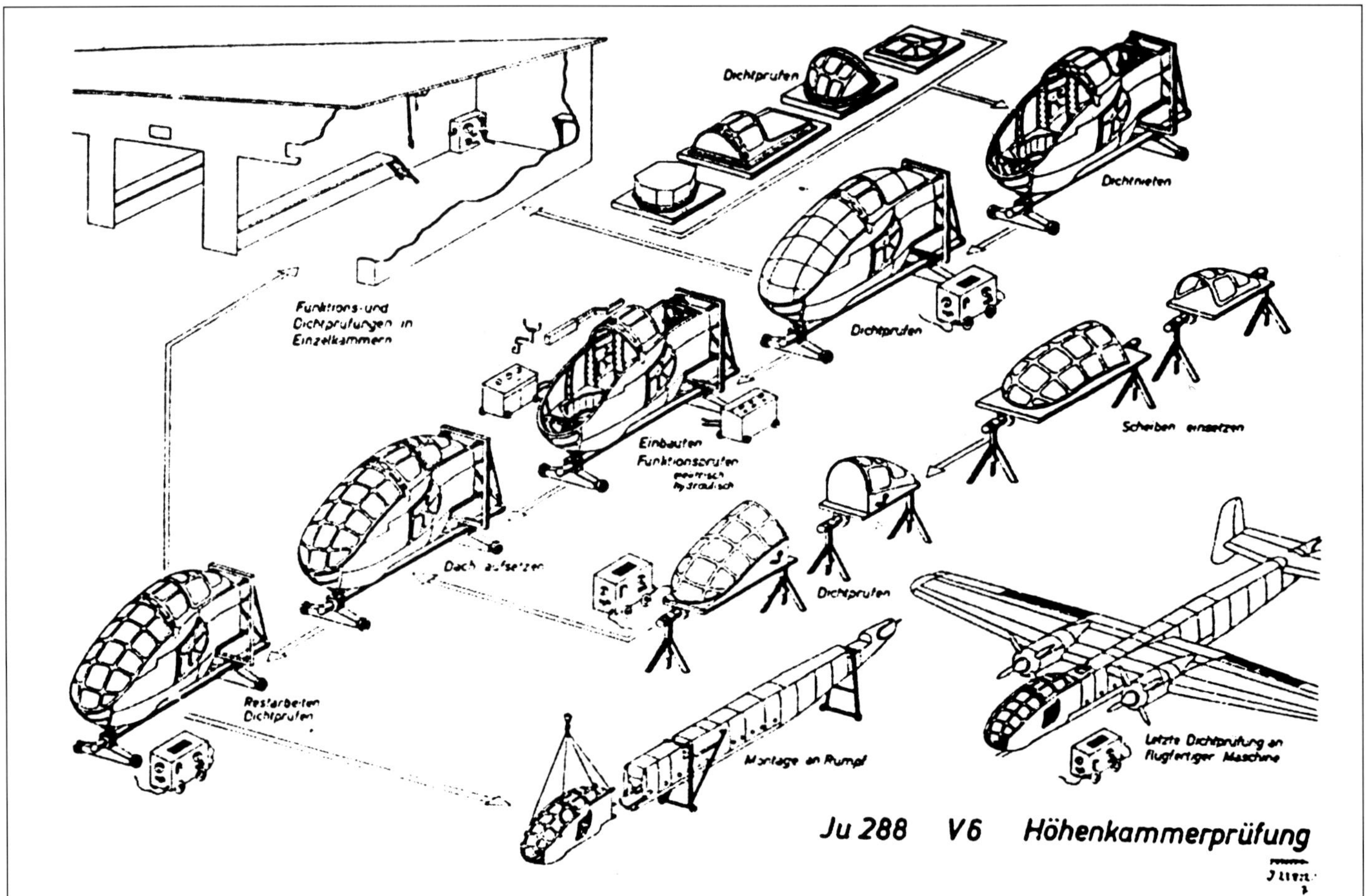

The assembly of the Ju 287 would probably have been similar to that of the Ju 288 shown here.

The equivalent types of drop canisters can also be accommodated. The bomb bay has been kept spacious enough so that hoisting and hanging of bombs can be easily accomplished. The main dimensions of the bomb bay are:

maximum length	5,471 mm
maximum width	1,660 mm
maximum height	980 mm

The Ju 287 is of all metal construction, with duraluminum and steel comprising the primary materials. In broad outline, the following are the approximate composition of the aircraft:

Metals:	duraluminum	62%
	steel	24%
	electron	4%
	other	0.5%
Non-metals		9.5%

The larger steel components are:

Undercarriage, bomb racks, beam girders in the wing center section including junction gussets to the wing girders, cockpit roof, engine mounts as well as a number of panels.

The wings are of a twin spar design with duraluminum girders constructed initially of extrusion press parts, later of forged press pieces. The wing center section consists of the two spars and an upper and lower shell. The leading edge section and the trailing edge section, which includes the control surfaces, are joined to the center section. The construction of the upper and lower shells as well as the trailing edge section is similar to that of the Ju 288. Handhole covers are planned at certain points to provide easy access to interior components.

Landing assistance in the form of a semi-articulated flap system is planned. The ailerons are equipped with semi-buried counterbalances, and are also partially extended when the landing flaps are set. For takeoff and landing, two positions are available:

	Landing Flaps	Ailerons
Takeoff:	30%	10%
Landing:	50%	16%

A safety measure is fitted against overstressing when the flaps are extended. The wing is equipped with an automatic extending slat along the inner wing sections. The span of this slat is about 1.3 m as measured from the fuselage sidewall.

Wing dihedral at the fuselage sidewall is 2%. Out to the beginning of the aileron the wing drops to -1.5% running in a straight line along the wing contours.

The retractable main undercarriage is housed in the inner wing section. The oleo and the retraction mechanism disappear into the wing wheel bay, which is fitted with a corresponding door.

The wing profile was determined using detailed calculations and wind tunnel measurements. The skeleton line has a curve of 1% to 23.3% chord and slopes upward to the rear to keep the zero moment of the profile at around zero. The greatest wing thickness is 40% that of the chord and totals 12.5% at the fuselage sidewall and 10.5% at the end rib. The wing profile is vertical in relation to the wing's plane of reference and is aligned in the direction of flight.

Profile identifier:	fuselage sidewall	1 23.3	12.5-0.825	40/0.178
	end rib	1 25.3	10.5-0.825	40/0.150

(The last figure is an identifier for the profile trailing edge)

The fuselage consists of four larger components: crew compartment, center section, aft section and fuselage end. The individual fuselage pieces are finished separately as independent components and then joined together to form the actual fuselage in the final assembly stage. The crew compartment is designed for high altitude operations; it consists of the frame construction and two roof sections. The four longerons in the crew compartment are riveted to four corresponding fuselage longerons.

The fuselage center section is completely taken up by the bomb bay in its lower part and by a three-section fuel compartment in its upper part. Construction of the center section consists of four longerons, four sickle-shaped frames, an interim floor between the bomb bay and fuel compartment, two bolted on side coverplates in the two forward fuel compartment sections and the two mechanized bomb bay doors that fold lengthwise when opened.

The aft section is constructed of two larger extruded duraluminum pressed pieces for the sidewalls, standard skin sections on the upper part and two undercarriage doors affixed to a cross member. RDF equipment is housed in the upper section. The duraluminum pressed parts absorb the movement of the wing forces into the fuselage.

The tail section houses two additional fuel tanks. Vertical and horizontal stabilizers are attached in the standard configuration. Construction consists of four longerons and the standard skin sections, with access hatches in various areas for maintenance and access to the fuel tanks and installed equipment. The fuselage tail terminates in the remote controlled tail barbette.

The empennage is, like the wings, of a twin spar design. The two horizontal stabilizer sections join together mid-fuselage, similar to the Ju 88. The two halves of the tailfin and the elevator are the same and are interchangeable. A symmetrical high speed profile was chosen for the horizontal and vertical stabilizers, with maximum thickness at 40% of the chord. For both control surfaces the thickness ratio is constant over the entire span and is 10% of the control surface chord. Identification data for the profiles is:

0 00 10-0.82540/0.125

In order to reduce the control surface forces, the horizontal and vertical stabilizers make use of internal counterbalances and trim tabs. For landing the elevators together with the landing flaps are hydraulically controlled.

The nose gear is hydraulically retractable. The main undercarriage retracts using a robust retraction strut that, when extended, acts as a support brace. The oleo strut retracts into the wing and the wheel into the fuselage center section. When retracting, the 1440x520 wheels rotate on an angled axle lying roughly in the direction of flight by means of a retraction cylinder inside the strut so that they are housed nearly vertically inside the fuselage. The main wheels are fitted with dual brakes. The nose gear pivots on an axle that is angled aft off the vertical by about 12%. A new type of braking system, designed to counter the wheel's flutter, is attached to the lower compression strut. This permits the wheel to rotate through a full 360% without the need for undoing any parts. Via a cam plate, two pistons move up and down inside two cylinders which, with a certain amount of overpressurized oil, causes them to be rocked back and forth by a damper. This damper effectively cancels any flutter effect. The dimensions of the wheel are 935x345. The nose gear is housed beneath the floorboard of the crew compartment and retracts backward into this area under the cockpit flooring by means of a retraction cylinder that acts as a brace when the gear is in the extended position.

The main and nose gears are equipped with a mechanism that uses the slipstream to rotate the tires, which saves wear and tear on the tires and improves oleo suspension during landing.

The fuel system consists of five fuel cells with a total capacity of 9,060 liters. Three identical tank, each with a capacity of 2,070 liters, are located in the fuselage center section; they are designed like the fourth fuselage tank and are installed through hatches in the port fuselage side. The third tank is fitted in place via the second tank's hatch and the removable bulkhead plate of the bulkhead 3. The fourth cell is installed between bulkheads 6 and 7. A fifth, unprotected tank of 1,650 liter capacity is located between bulkheads 7 and 8 and is installed from below via a hatch in the fuselage belly. All tanks drain from the bottom valves and are equipped with electrical fuel supply gauges. The forward four cells are installed at the same level. The drain tanks gravity-feed their contents via a shared fuel line to a pump tank located ahead of bulkhead 5 at the lowest point of the system. This tank supplies the six engines at an equal rate. Of the eight tank pumps installed in the pump tank, two each are powered by four different circuits. To avoid the aircraft from becoming overtrimmed by shifting fuel in the feeder tanks at changing angles of attack, the tanks are fitted with back-pressure valves with minimum throughflow resistance at their drain points. These valves prevent any backflow of fuel into the tanks. This is important, for example, if one of the tanks is hit by fire and the fuel wants to flow from the line into the unsealed tanks. The feed line to the pump tank is located in a protected spot beneath the tank flooring between the bomb racks and is made of steel. The pump tank is armor protected against enemy fire and impact on belly landing.

With regard to the center of gravity, tank 5 is the first to drain. A hydraulically operated float valve is located inside its feed line. The tank's contents are fed to the pump tank by means of manifold air pressure. Once tank five has successfully emptied, fuel feed continues uninterrupted from the remaining four forward feeder tanks. To this end, the pump tank is equipped with exhaust valves that direct the inflowing manifold air into the collective exhaust line. A little while after the feeder tank shows empty the float valve closes, sealing off the flow of manifold air to the pump tank. Additionally, due to center of gravity constraints the feeder tank must be the last to drain. For this reason a vent line runs from the base of tank 3 to the top of tank 2, so that the air is only able to flow from tank 3 to tank 2 once the former has successfully drained, at which point fuel from the latter can begin flowing.

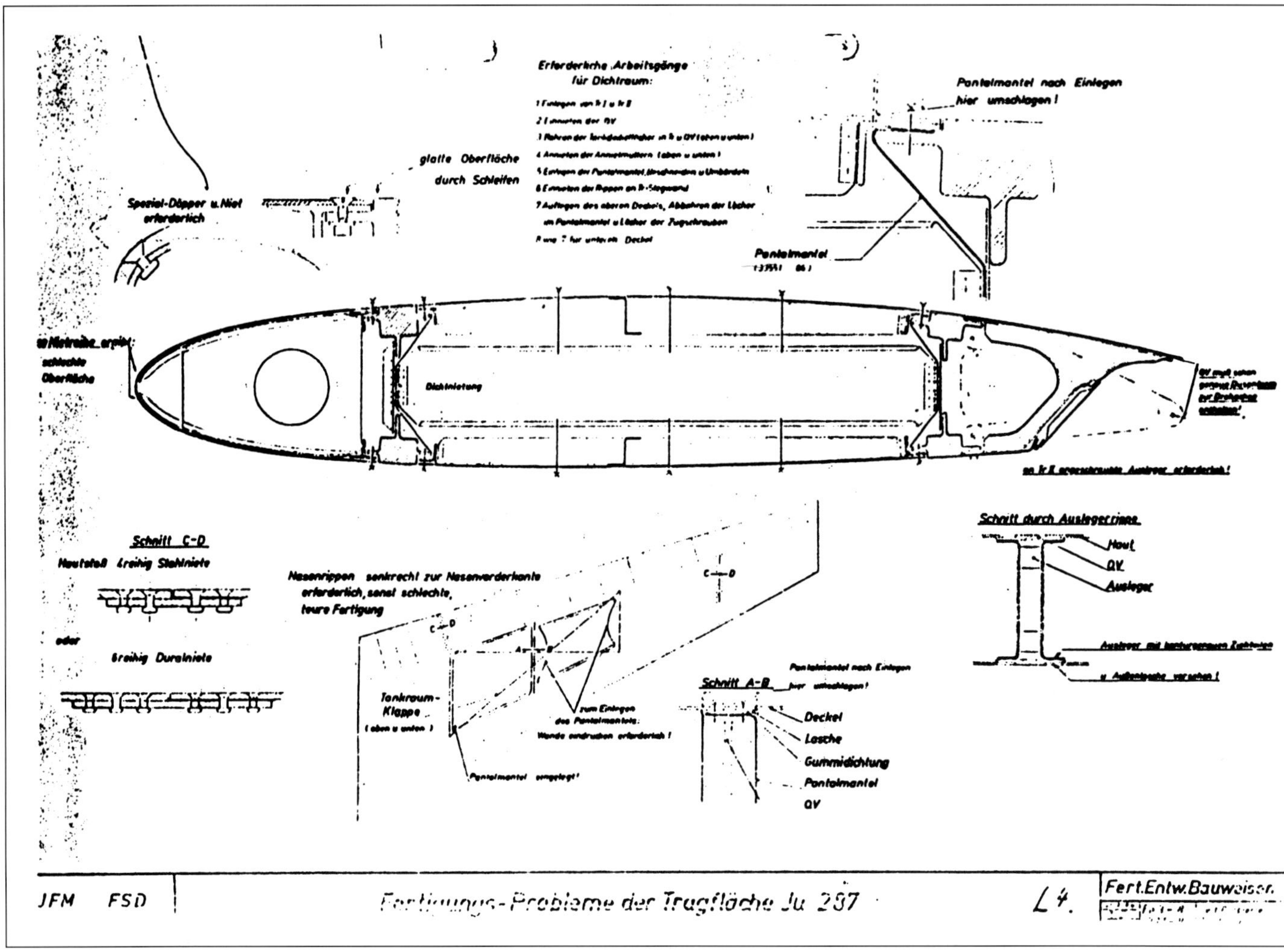

Factory blueprints clearly show the three-piece wing assembly.

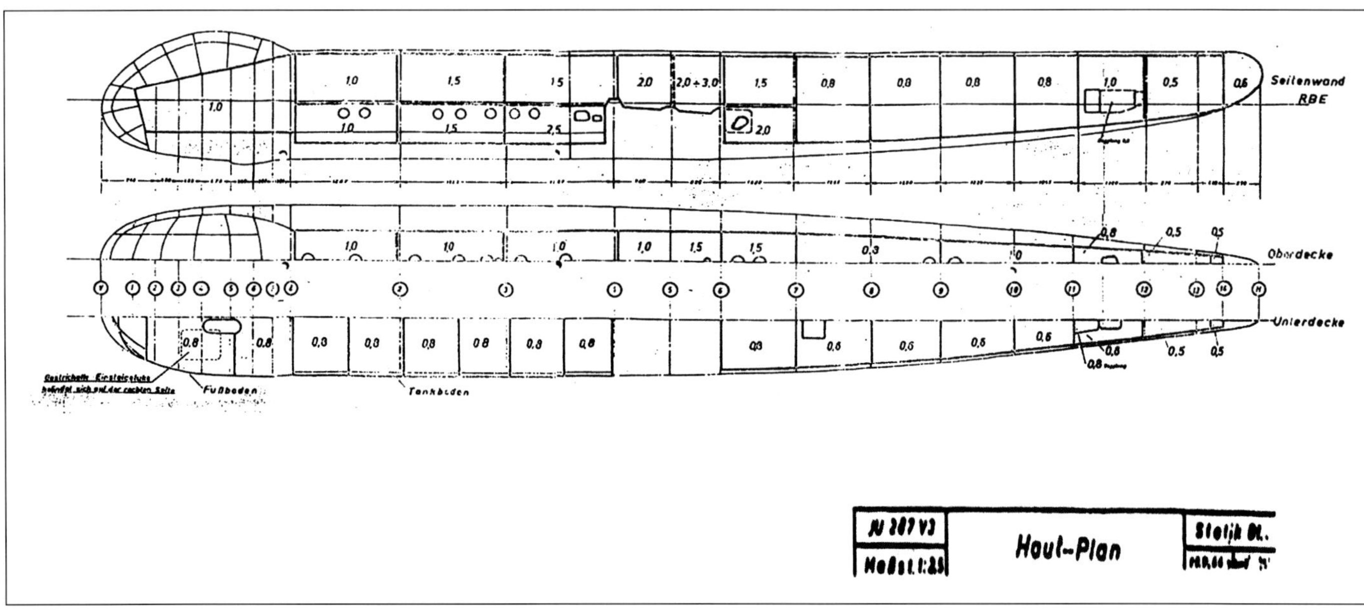

The so-called "skinning" blueprints show the skin sections for the fuselage.

The tanks are fueled from above via the forward feeder tank and the feeder line. To accomplish this, the back-pressure valves and the block vent at tank 5 must first be opened, which is done simultaneously using a Bowden cable. The back-pressure valves return to their original position after fueling has been completed. In addition, rapid fueling under pressure is accommodated by the planned incorporation of a single point receptacle in the feeder line.

The fuel system is equipped with a fuel dump, and the tanks can be pumped dry using manifold pressure air. The fuel dump valve is designed in such a way that the release of fuel can be interrupted at any time. The dump valve is closed by a hand crank located inside the crew compartment. The system vents the fuel through a tube extending from the fuselage belly between bulkheads 8 and 9.

The four forward main feeder tanks are designed as 4 mm thick fuel bladders that offer minimal yet effective protection against enemy fire. The effectiveness of this light tank protection is enhanced by a fire extinguisher system developed by Junkers.

Control is normal and requires no special measures. To maintain control over the control surface forces the aileron and rudder controls have been designed as spring controlled surfaces. The forces on the elevator controls are minimized through the use of direct-set trim tabs. In addition, plans call for a 2:1 reduction gearing, selectable during flight, for the elevator controls in order to not

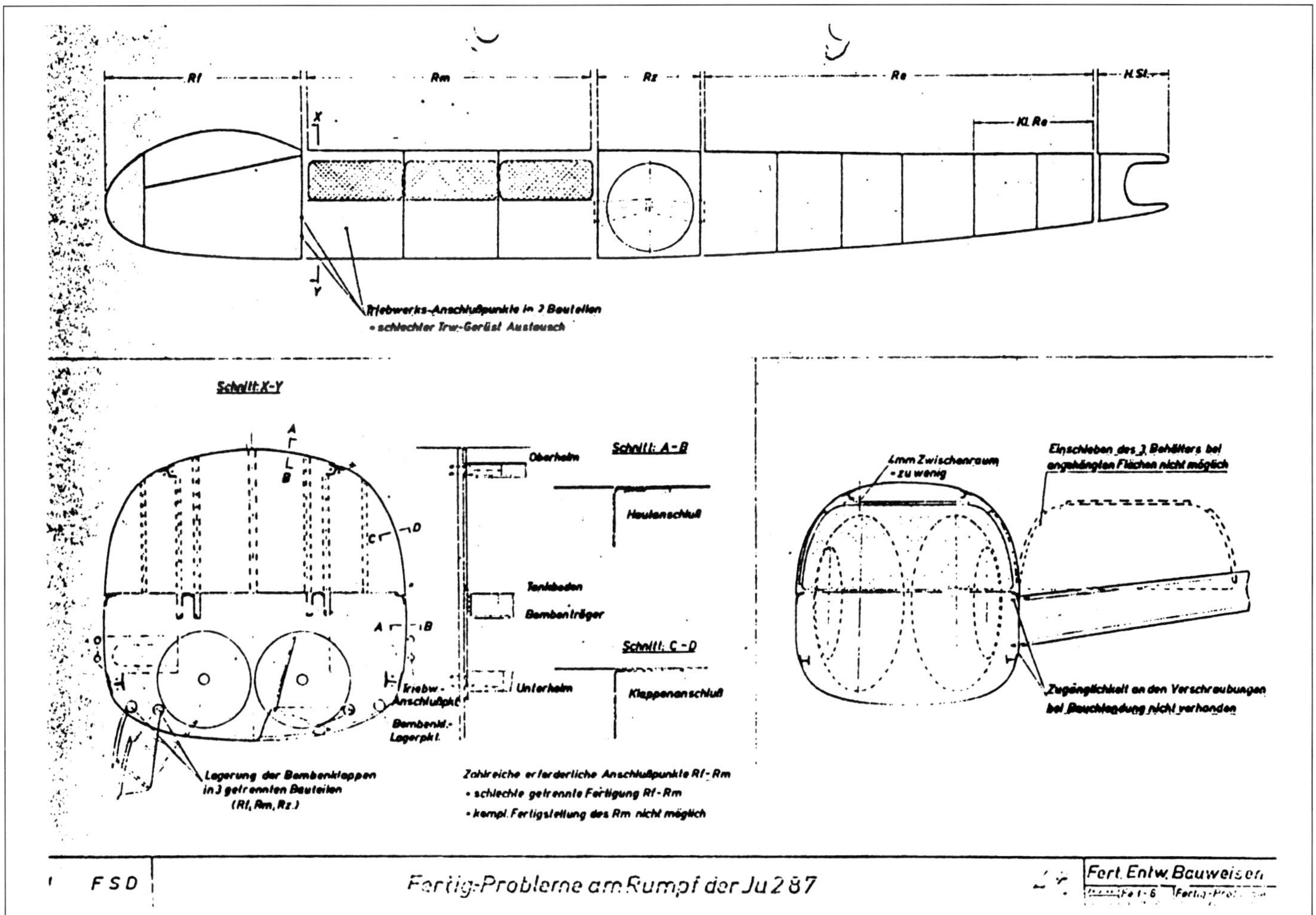

This factory drawing shows the problems inherent in the originally planned installation of the two fuselage-mounted engines. Other features include the planned arrangement of the bomb bay doors as well as the main undercarriage in its retracted position.

only adjust the control surface forces to the dynamic pressure , but also the hand controls as well. The entire control system is designed mainly from individual components and proven parts from other types.

The equipment basically corresponds to the standard equipment found on medium level bombers. For the radio operator, the PVE 11 periscope sighting mechanism, needed for operating the tail guns, protrudes slightly from the fuselage. At bulkhead 2 in the bomb bay there are six oxygen bottles located beneath the flooring, each with a 2 liter capacity for the three man crew using full oxygen throughout the entire flight. Three 10 liter bottles filled with dichlorine are located beneath the cockpit flooring for the fire extinguisher system. For takeoffs with heavy bomb loads and at higher external temperatures, the fuselage tail underside between bulkheads 10 and 11 is fitted with attachment points for three WASAG solid fuel rockets, each with 1,000 kg thrust with a burn time of 12 seconds. The expended casings are jettisoned after takeoff. Also located on the fuselage tail underside between bulkheads 11 and 12 is the planned housing for a braking parachute, which would accept a 6.1 meter diameter parachute. It is activated from the cockpit, and in emergency situations the deployed parachute can be detached.

Radio equipment is determined by penetration range, altitude and role, with the following equipment intended as standard:

Fu G 16Z (y)
Fu Bl 2K
Fu G 25 K
Fu G 136
Fu G 123

The following are additional planned options:

Fu G 217
Fu G 104
Fu G Berlin
Fu G 10
Emergency direction receiver

For the long-range reconnaissance variant, special equipment includes three RB 75/30 aerial cameras housed in the bomb bay. One of these is vertically mounted and the two others are angled 20 degrees to the right and left, respectively. Another possibility for an oblique installation of the two angled cameras is at an angle of 60 degrees. The remainder of the bomb bay is taken up by an additional 2,000 liter fuel tank. In the tail, it is possible to install either an RB 50/30 or a 50/18 camera between bulkheads 8 and 9. For the reconnaissance version, the bomb bay doors will be replaced by a new belly of wooden construction.

For a night reconnaissance platform version, two RB 35/25 cameras are installed in the bomb bay, angled at about one-half field of view (about 18 degrees) to either side. 16 flares are carried on standard bomb racks in the aft section of the bomb bay. The remainder of the bomb bay serves as space for an auxiliary fuel tank of 1,200 liter capacity. On the night reconnaissance variant the bomb bay doors remain and are opened when photos are being taken.

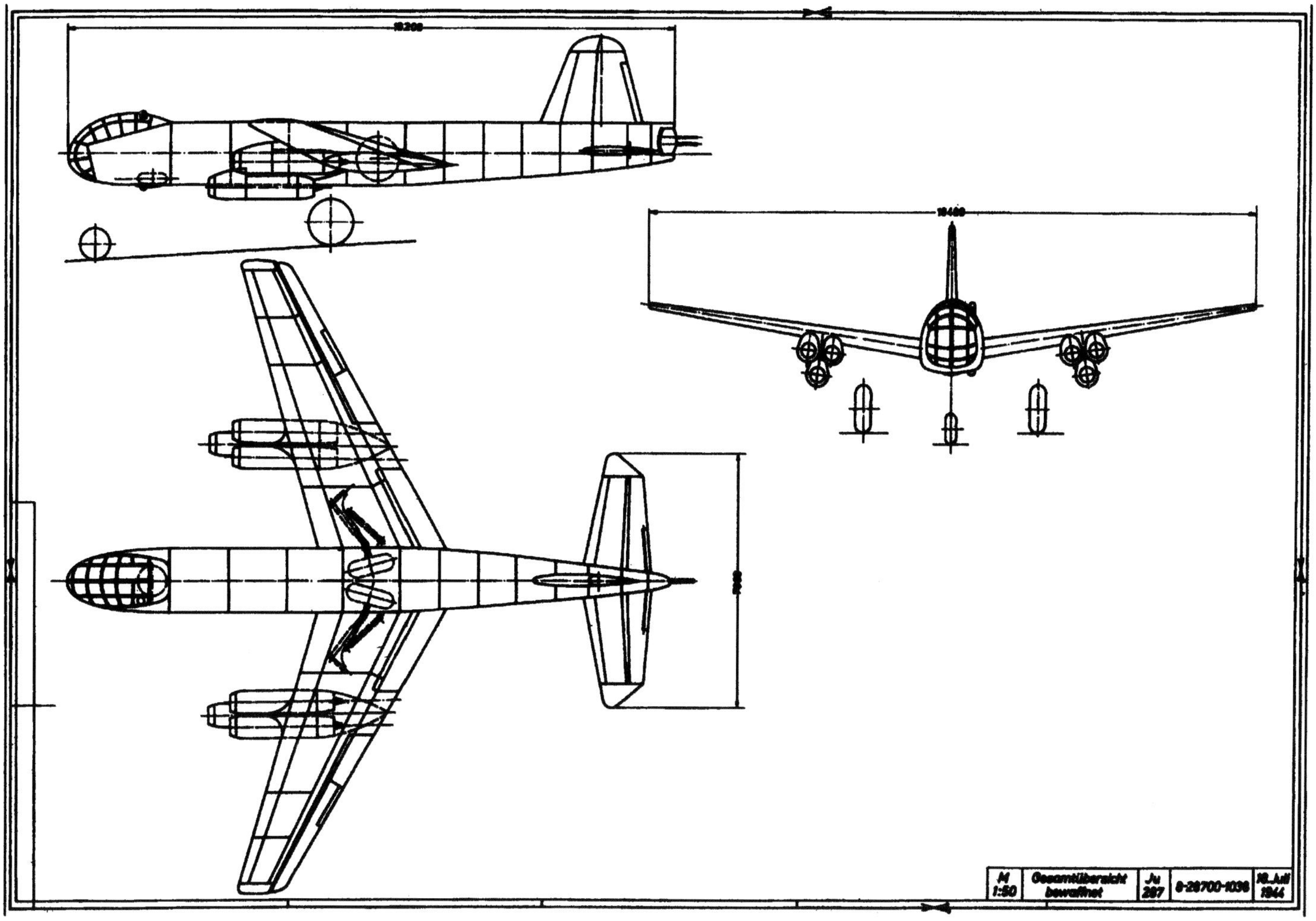

Above: Three-view of the Junkers Ju 287 as intended for full-scale production, with simplified canopy glazing.

Front view of the cockpit mockup with refined canopy for improved forward downward visibility.

Side view of the mockup with the modified "nose glazing".

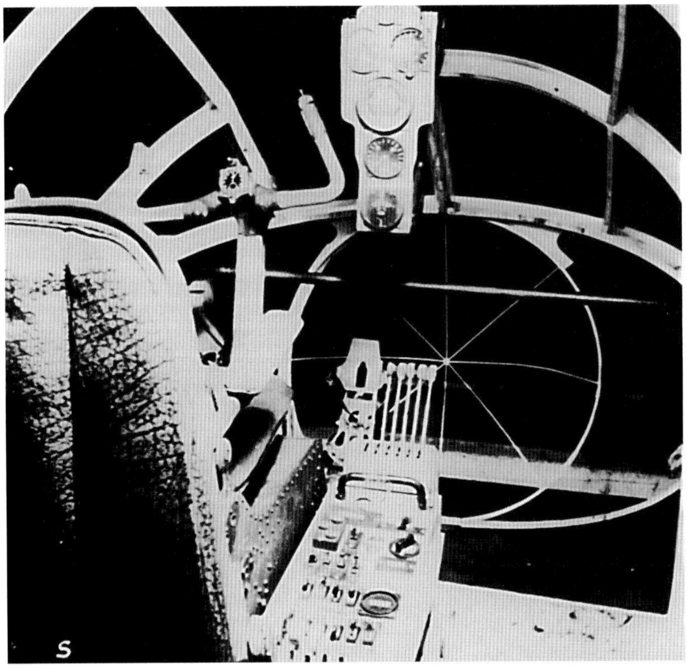

This photo, viewing the forward crew compartment, clearly shows the better visibility afforded to the pilots.

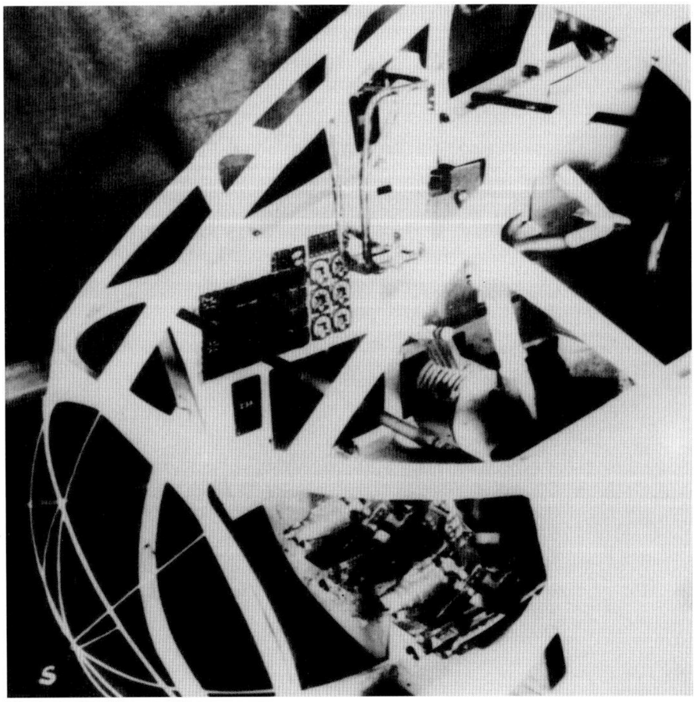

And because it looks so nice, yet another shot of the cockpit interior taken from the other side. It shows a well-stocked instrument bay located on a side console on the starboard inner instrument panel.

The bare workspace for the radio operator and gunner.

Below: When finished, the work area would most likely have looked similar to this, with the periscope sight on the right. The picture shows the same layout with PVE 11 sight in a Ju 388.

FHL 131 Z tail turret as originally intended for the Ju 287, here seen during installation in a Ju 388 K. The bottom MG 131 has not yet been fitted.

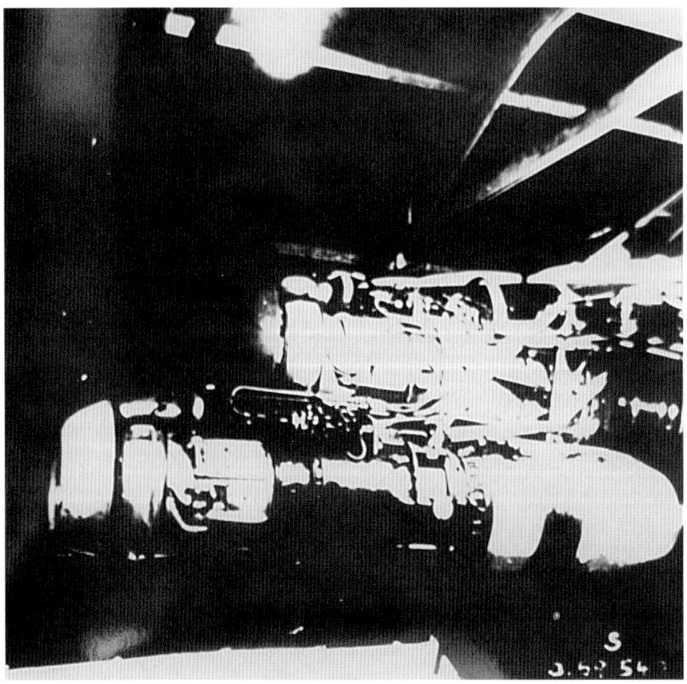

BMW tripe engine configuration mounted to the wing for test runups.

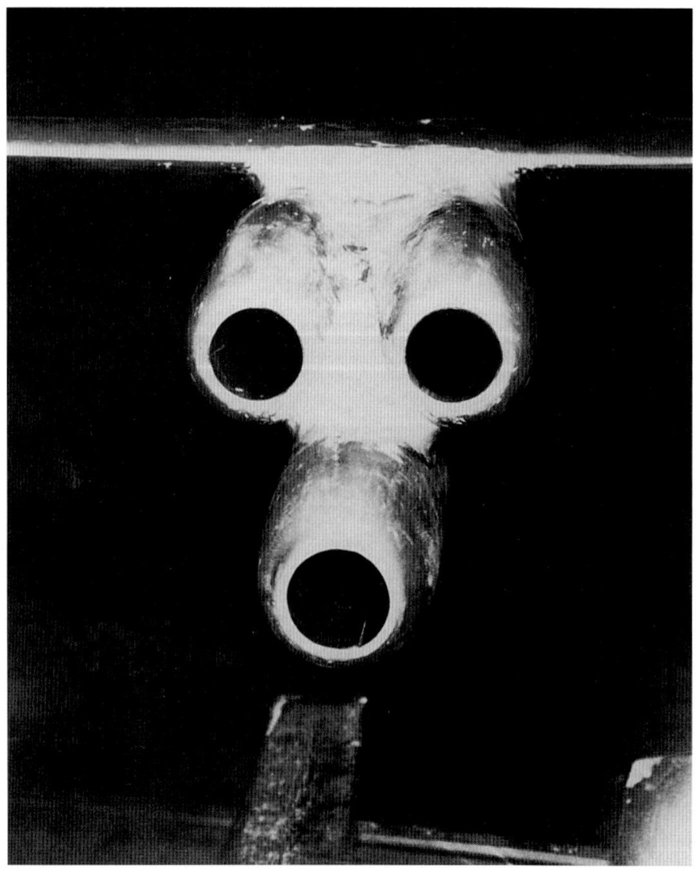

Model of the engine pods in triple configuration. The shape of the engine nacelles was modeled using plasticine.

Junkers Flugzeug- und Motorenwerke A.G. Flugzeugbau Stammwerk Dessau	Bearbeitung Kobü-Entwurf Dr. Cr./Mix/MS	Datum 7.9.44	Kennzeichnung Kobü-Ent. 4549

Bl. -10 -

Geheime Kommandosache!

Errechnete Leistungen Ju 287

	Triebwerk		6 x BMW 003 A1
	Flügelfläche	m^2	58,4
	Spannweite	m	19,4
	Bombenzuladung	kg	3000
	Rüstgewicht Besatzung Kraftstoff Munition Bomben	kg " " " "	12180 300 7200 100 3000
	Abfluggewicht	kg	22780
6-motorig	v_{max} mit G_{mittel} in H = 0 H = 6	km/h "	820 870
6-motorig	Errechnete Flugstrecke in H = 0 H = 10-12	km km	740 1950
6-motorig	Steiggeschw. mit $G_{Abfl.}$ in H = 0 " G_m " H = 6 " G_m " H = 10	m/s " "	12,3 10,4 4,2
6-motorig	Dienstgipfelhöhe $G_{Abflg.}$ -Steigkraft- G_m stoff	km "	10,3 11,8
6-motorig	Steigzeiten mit $G_{Abflg.}$ bis H = 10 km	min	26,5(9,7)
bei Ausf. v. Triebw.	v bei Ausfall von 1 Triebwerk (Flugverlauf siehe Schaubild)	km/h	710
bei Ausf. v. Triebw.	Flugstrecke mit 5 Triebwerken beim Triebwerksausfall beim Bombenwurf	km	1830
bei Ausf. v. Triebw.	Steiggeschw. bei $G_{Abflg.}$ in H = 0 bei Ausf. von 1 Triebwerk Ausf. von 2 Triebwerken	 m/s m/s	 7,3 3,5
bei Ausf. v. Triebw.	Dienstgipfelhöhe bei Ausfall von 1 Triebw. mit G_m Ausfall von 2 Triebw. mit G_m	 km km	 9,8 7,3
	Rollstrecke (o.Zus.-Schub)	m	1500
	Steiggeschwindigkeit unmittelbar nach Abheben bei ausgefahr. Fahrw. u. Klappen bestes Steigen nach Start	m/s m/s	4,2 4,8
	Landegeschwindigkeit bei Landegewicht (1500 kg Kraftstoff)		176 14000
	Ausrollstr.b.norm.Platzverh.o.Bremsschirm b.ungünstigen " m. "		480 600

Performance chart for the Ju 287 with six BMW 003 A1 turbojet engines.

Flugzeugbau Sommerwerk Dessau | Tr.Gr./[illegible] | 6.9.44 | Kobü-Nr. 4549

Bl. -11 -

Geheime Kommandosache!

Errechnete Leistungen Ju 287

Triebwerk		6 x Jumo 004 - B
Flügelfläche	m^2	58,4
Spannweite	m	19,4
Bombenzuladung	kg	2000
Rüstgewicht Besatzung Kraftstoff Munition Bomben	kg " " " 	13590 300 7200 100 2000
Fluggewicht	kg	23140
v_{max} mit G_{mittel} in H = 0 H = 6	km/h "	850 870
Errechnete Flugstrecke in H = 0 H = 10 - 12	km "	700 1700
Steiggeschw. mit $G_{Abfl.}$ in H = 0 " G_m in H = 6 " G_m in H = 10	m/s " "	15,4 11,8 3,50
Dienstgipfelhöhe $G_{Abfl.}$ -Steigkraftstoff G_m	km "	10 11
Steigzeiten m. $G_{Abfl.}$ bis H = 10 km	min	14 (7,8)
v bei Ausfall von 1 Triebwerk (Flugverlauf siehe Schaubild)	km/h	650
Flugstrecke mit 5 Triebwerken beim Triebwerksausfall beim Bombenwurf	km	1600
Steiggeschw. bei $G_{Abfl.}$ in H = 0 bei Ausf. von 1 Triebwerk Ausf. von 2 Triebwerken	 m/s "	 9,90 5,70
Dienstgipfelhöhe bei Ausfall von 1. Triebw. mit G_m Ausfall von 2 Triebw. mit G_m	 km "	 9,55 7,50
Rollstrecke	m	1600 (6x900)
Steiggeschwindigkeit unmittelbar nach Abheben bei ausgefahr. Fahrw. u. Klappen bestes Steigen nach Start b.ausg.Fw.u.Kl.	m/s "	4,10 4,70
Landegeschwindigkeit bei Landegewicht (1500 kg Kraftstoff)	km/h kg	183 15200
Ausrollstr.b.norm.Platzverh.o.Bremsschirm " b.ungünstigsten "	m m	520 670

Performance chart for the Ju 287 with six Jumo 004 B turbojet engines.

Junkers Flugzeug- und Motorenwerke AG Flugzeugbau Stammwerk Dessau	Bearbeitung Kobü-Entwurf Dr. Cr./Mix/Si.	Datum 6.9.44	Kennzeichnung Kobü-Sw. 4549

Geheime Kommandosache! Bl. -12-

Errechnete Leistungen Ju 287

	Triebwerk		6 x Jumo 004 - C
	Flügelfläche	m^2	58,4
	Spannweite	m	19,4
	Bombenzuladung	kg	3000
	Rüstgewicht Besatzung Kraftstoff Munition Bomben	kg " " " "	13540 360 8200 100 3000
	Abfluggewicht	kg	25130
6-motorig	v_{max} mit G_{mittel} in H = 0 H = 6	km/h "	890 905
6-motorig	Errechnete Flugstrecke in H = 0 H = 10 - 12	km "	760 1930
6-motorig	Steiggeschw. mit $G_{Abfl.}$ in H = 0 " G_m in H = 6 " G_m in H = 10	m/s " "	16,8 14,8 5,7
6-motorig	Dienstgipfelhöhe $G_{Abfl.}$-Steigkraftstoff G_m	km "	10,3 11,7
6-motorig	Steigzeiten m. $G_{Abfl.}$ bis H = 10 km	min	14 (8,20)
bei Ausf. v. Triebwerken	v bei Ausfall von 1 Triebwerk (Flugverlauf siehe Schaubild)	km/h	670
bei Ausf. v. Triebwerken	Flugstrecke mit 5 Triebwerken beim Triebwerksausfall beim Bombenwurf	km	1820
bei Ausf. v. Triebwerken	Steiggeschw. bei $G_{Abfl.}$ in H = 0 bei Ausf. von 1 Triebwerk Ausf. von 2 Triebwerken	m/s "	10,80 6,0
bei Ausf. v. Triebwerken	Dienstgipfelhöhe bei Ausfall von 1 Triebw. mit G_m Ausfall von 2 Triebw. mit G_m	km "	9,90 8,0
	Rollstrecke	m	1650 (6x1015)
	Steiggeschwindigkeit unmittelbar nach Abheben bei ausgefahr. Fahrw. u. Klappen bestes Steigen nach Start b.ausg.Fw.u.Kl.	m/s "	4,50 5,20
	Landegeschwindigkeit bei Landegewicht (1500 kg Kraftstoff)	km/h kg	183 15150
	Ausrollstr. b.norm.Platzverh.o.Bremsschirm	m	515

Performance chart for the Ju 287 with six Jumo 004 C turbojet engines.

To conclude this chapter, another incomparable model photograph by Günter Sengfelder. The picture shows a combat version of the Ju 287 at a fictional airfield.

So much for the original description. It is worth noting that many of the details were actually implemented in simplified form. One interesting example worth mentioning is that prior to touchdown on landing the wheels were set spinning in order to keep wear on the tread to a minimum. This was accomplished by placing a cylindrical fan on either side of each wheel, similar to what can be found nowadays in a more elongated form on night storage heating units. The whole thing looked something like a water wheel, but with the difference that the airflow would set the fan rotating, which in turn would start the wheels spinning.

Also of particular note is certainly the fact that a comprehensive and apparently quite effective fire extinguishing system was installed. Which must have been an absolute necessity for the engines at the time, which were often quite troublesome and easily prone to catching fire.

The Ju 287's superior performance, which is shown on pages 105-107 with several possible engine configurations, could only be achieved by making use of the jet engines that were at that point still in their teething stages. For this reason the next chapter will examine the two most important engine types that would play a role in the brief life of the Ju 287 - the BMW 109-003 and the Jumo 109-004 turbojet engines.

The BMW 109-003 and Junkers Jumo 109-004 Jet Engines

This chapter is not intended to be a detailed technical treatment of jet engines; its purpose is to offer brief, succinct sections outlining the key points of what was then a revolutionary propulsion system and the associated problems inherent in any new technology.

With the end of the 1930s, speeds exceeding 700 km/h had been successfully logged, and it quickly became apparent that any further increase in speed using propeller propulsion was fast becoming impossible. New alternatives had to be found. As early as the beginning of that decade some persistent folks had tinkered around with the widest variety of reaction-type propulsion systems, but no one really took them seriously. Then, however, the situation changed when Ernst Heinkel, in a 1939 company funded project, successfully launched the rocket-powered He 176 and the turbojet-powered He 178, both of which were eyed suspiciously by the RLM.. Pioneer efforts, although proper recognition for these deeds were initially long in coming. The RLM offered courteous congratulations, to be sure, but the consequences of these flights - particularly the first turbojet powered flight in the world - were not readily apparent at first. Nevertheless, it was a start.

Curiously, the RLM then issued a contract in 1939 for development of a turbojet engine with a static thrust of 600 kg (at that time thrust was still measured in kg, although this later changed to kp). However, the contract did not go to Heinkel, but to BMW. Ten experimental engines were to be developed for use in various tests.

Now, of all the potential companies, why did the RLM award the developmental contract to the BMW company? The question can be answered easily and quickly. *Dr.-Ing.* Hermann Oestrich had already been working for years on the development of a functional turbojet engine, with some success in this area already. Unlike the Heinkel turbine, which used a so called radial compressor, the BMW experimental turbine employed an axial compressor, which offered the promise of much higher performance with future developments. Additionally, the diameter of the axial compressor could be kept much smaller than the radial compressor turbine (see picture p. 110, which clearly shows the difference between the radial and the axial compressors).

The engine was assigned the project number P 3302. Problems constantly had to be wrestled with. Everything, truly everything, was new territory: the shape of the intake cover, the compressor, the combustion chamber, the turbine, the exhaust nozzle, etc. There seemed to be no end to the difficulties, and the years rolled along. In 1941 the engine provided an unacceptable 150 kg thrust during bench testing. The main problem was discovered to be with the manufacturing of the turbine blades. As 1941 turned into 1942, it became obvious that the once hoped-for goal could not be achieved, and another experimental engine series, the V11-V14 was established with new design criteria. The finalized criteria for the V1-V10 series had only offered a maximum possible static thrust of 550 kg. Work on the new test prototypes began at the end of 1942. During testing, the blades on the first stage of the compressor suffered regular breakage, and the axial thrust bearings on the compressor often became chewed up. Yet by and by the worst of the problems were corrected, so that by 1943 a pre-production series with the RLM designation of 109-003 A-0 was set.

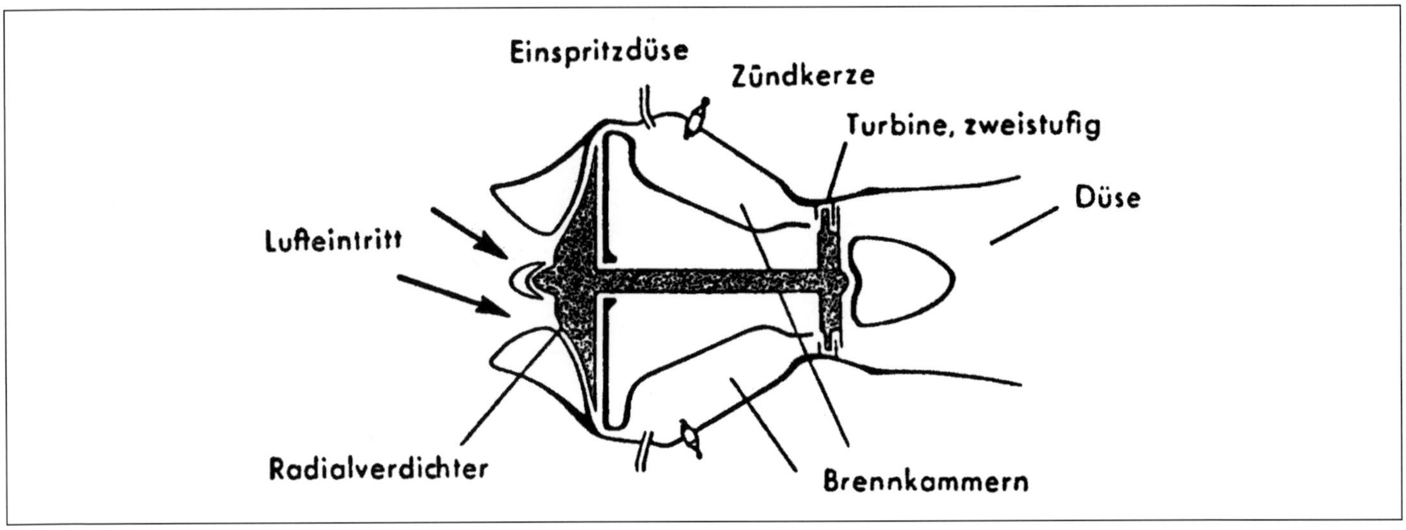

The radial turbine is characterized by a large-diameter turbine compressor fan and functions on the principle of a centrifugal blower. Although relatively simple to produce, it takes up larger space and is quite limited in its performance potential.
Key:
Lufteintritt - air intake
Radialverdichter - radial compressor
Brennkammern - combustion chambers
Düse - exhaust nozzle
Turbine, zweistufig - two-stage turbine
Zündkerze - ignitor
Einspritzdüse - fuel nozzle

It was able to provide a constant 600 kg thrust without difficulty, with a service life of twenty, then later up to fifty hours.

Through additional detail improvements thrust was increased to 800 kg and operational reliability was constantly improved. This engine from the BMW company went into production as the BMW 109-003 A-1 and was installed in at least some versions of all turbojet powered aircraft, such as the Arado Ar 234, the Heinkel He 162, the Junkers Ju 287 (V2 and V3) and the Messerschmitt Me 262.

By August 1944 1,000 production engines had been delivered. The ever increasing hardships of war forced a switch from the B4 fuel (standard aircraft fuel) used up to that point to a diesel-like J2, which was still available in sufficient quantities. The use of this J2 fuel initially negatively affected the startup qualities of the jet engines. However, by modifying the starter system the problem was quickly brought under control.

In order to improve operational reliability even further, the engines underwent yet another major rework and some units were

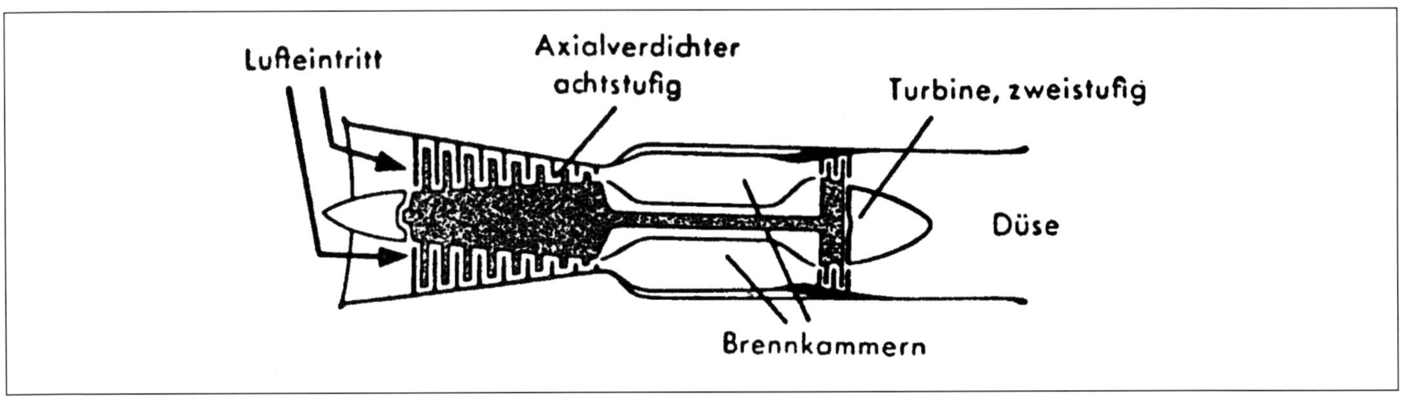

The axial turbine uses several compressor blades arranged in tandem along an axis and, despite a smaller space needed for installation, has a relatively higher performance potential.
Key:
Lufteintritt - air intake
Axialverdichter achtstufig - eight-stage axial compressor
Brennkammern - combustion chambers
Turbine, zweistufig - two-stage turbine
Düse - exhaust nozzle

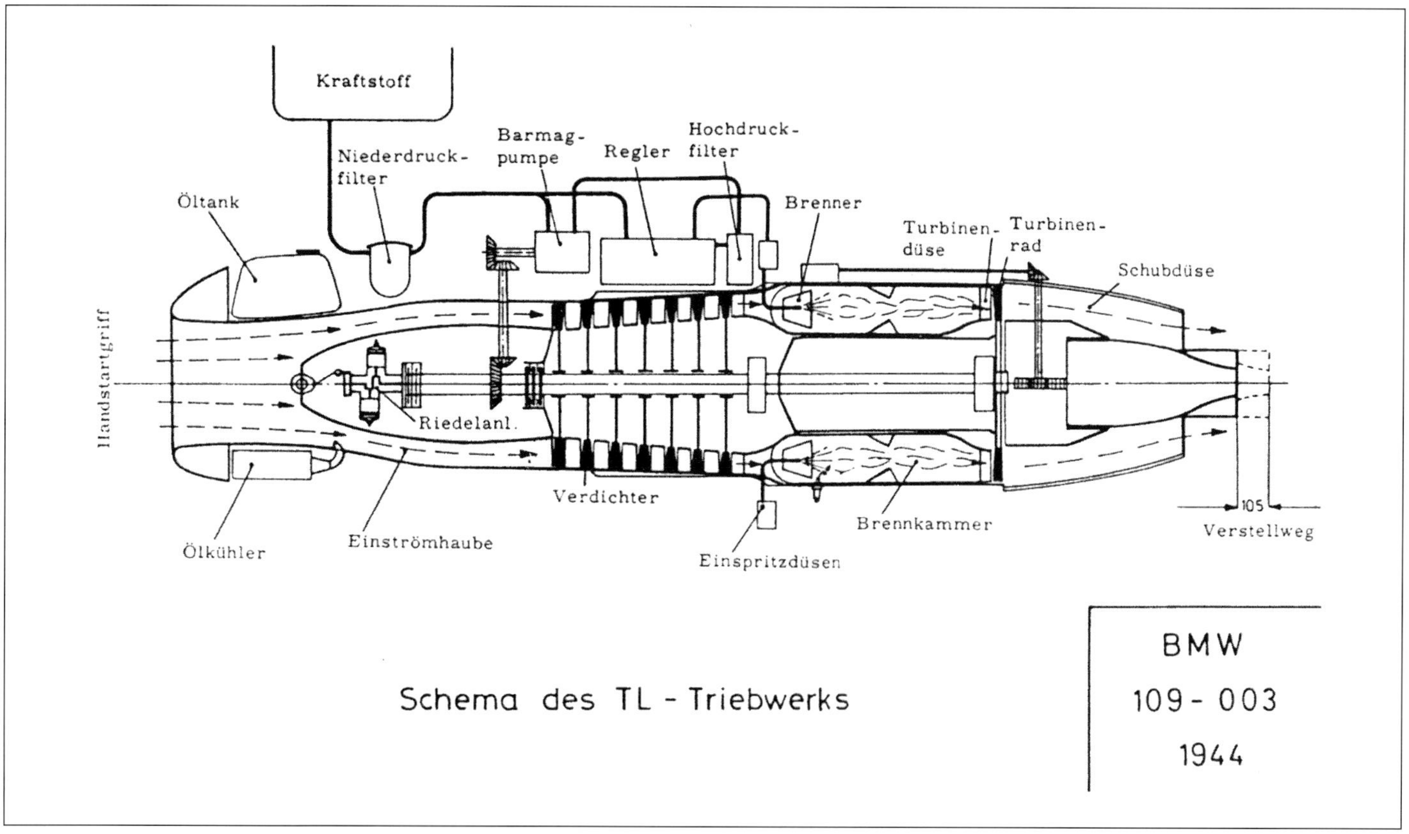

Cutaway of the BMW 109-003 turbojet engine. Compared to the Junkers engine, the internal combustion (or Riedel) starter is set far deeper inside the engine opening, a feature that easily distinguishes the two.

Key:

Handstartgriff - starter handle
Öltank - oil tank
Kraftstoff - fuel
Niederdruckfilter - low pressure filter
Barmagpumpe - barmag pump
Regler - governor
Hochdruckfilter - high pressure filter
Brenner - combustor flame
Turbinendüse -turbine nozzle
Turbinenrad - turbine
Schubdüse - exhaust nozzle
Verstellweg - travel distance
Brennkammer - combustion chamber
Einspritzdüsen - fuel nozzles
Verdichter - compressors
Einströmhaube - intake duct
Riedelanl. - Riedel starter
Ölkühler - oil radiator

able to receive this engine, designated the BMW 109-003 A2, shortly before the end of the war.

In addition to the BMW turbojet engine, an in-house design - the Jumo 109-004 - would also be used on the Ju 287. But this is its own story, as we shall shortly see.

In 1937 a small group under the direction of Professor Herbert Wagner and Max A. Müller privately began working on a jet engine at Junkers' Magdeburg branch. Just like BMW, a turbojet engine with an axial compressor was favored. The results they produced, however, were not particularly encouraging and the team of colleagues disbanded despondently. Ironically, Max A. Müller transferred jobs over to Heinkel, a strong competitor of Junkers. Yet despite all the failures Junkers was also awarded a contract by the RLM for the development of a jet engine parallel to that of Heinkel. The contract was given to Professor Otto Mader, who passed the project management on to *Dr.-Ing* Anselm Franz. He reorganized the project from the ground up. Actual construction began in October 1939 under the supervision of Fritz Böttger. The first test runup of this T1 engine, which proved to be quite successful, occurred on 11 October 1940. Not surprisingly, development brought along with it a few problems, to be sure, but all in all was remarkably successful. As early as August 1941 the required static thrust of 600 kg was reached, and in January 1942 1,000 kg thrust was even produced. Things happened quickly from this point onward. The Junkers turbojet was taken into the RLM's supply program as no. 109-004. On 18 July 1942 the Messerschmitt Me 262 V3 made its first pure jet-powered flight with two experimental versions of the Jumo 109-004. Production began in October 1943 with a rate of up to 1000 units per month being achieved shortly before the war's end. All told, over 6000 production engines were built. The Junkers Jumo 109-004 was, despite its teething troubles, the first turbojet engine in the world to go into full-scale production and proved itself admirably in all airframes using this new type of propulsion. Compared to the BMW engine the last standard version of the Jumo, the 109-004 B1, was about 100 kg more powerful with a thrust of 900 kg. This leads to the question: why was Junkers suddenly able

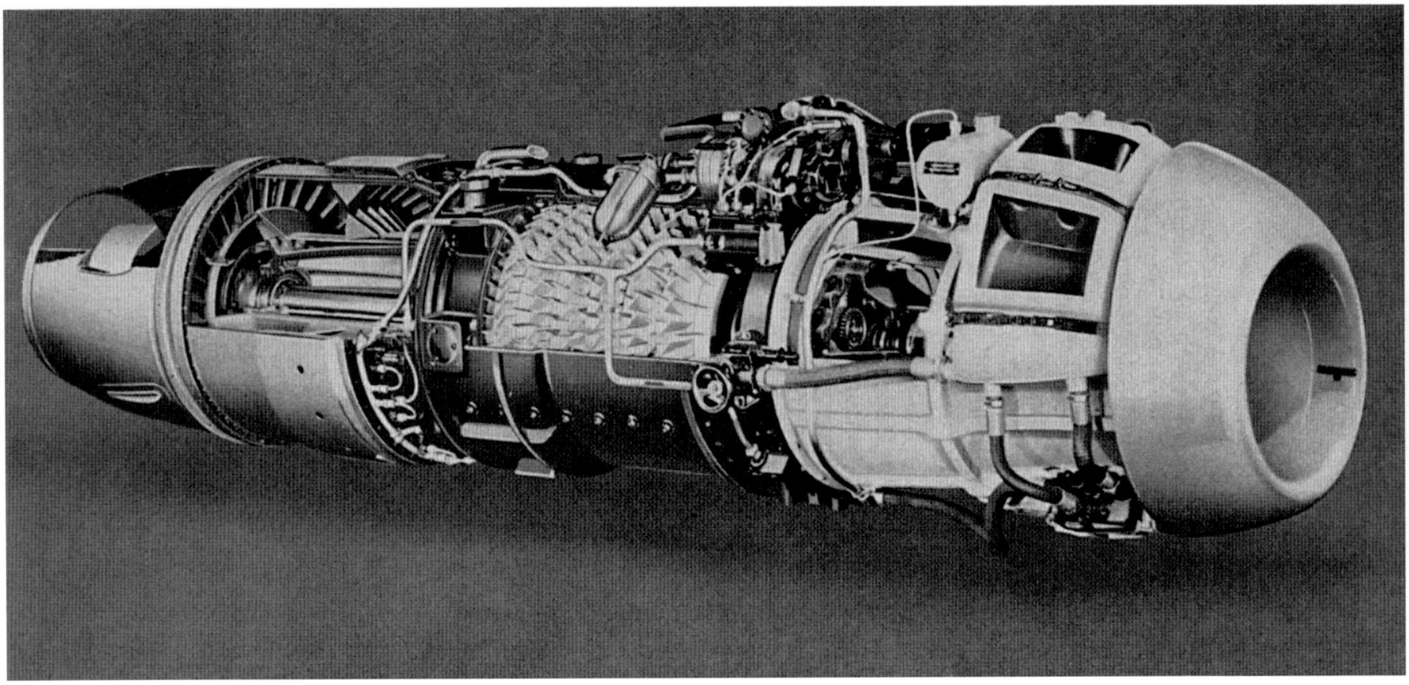

A BMW 109-003 A1 engine with cover panels removed.

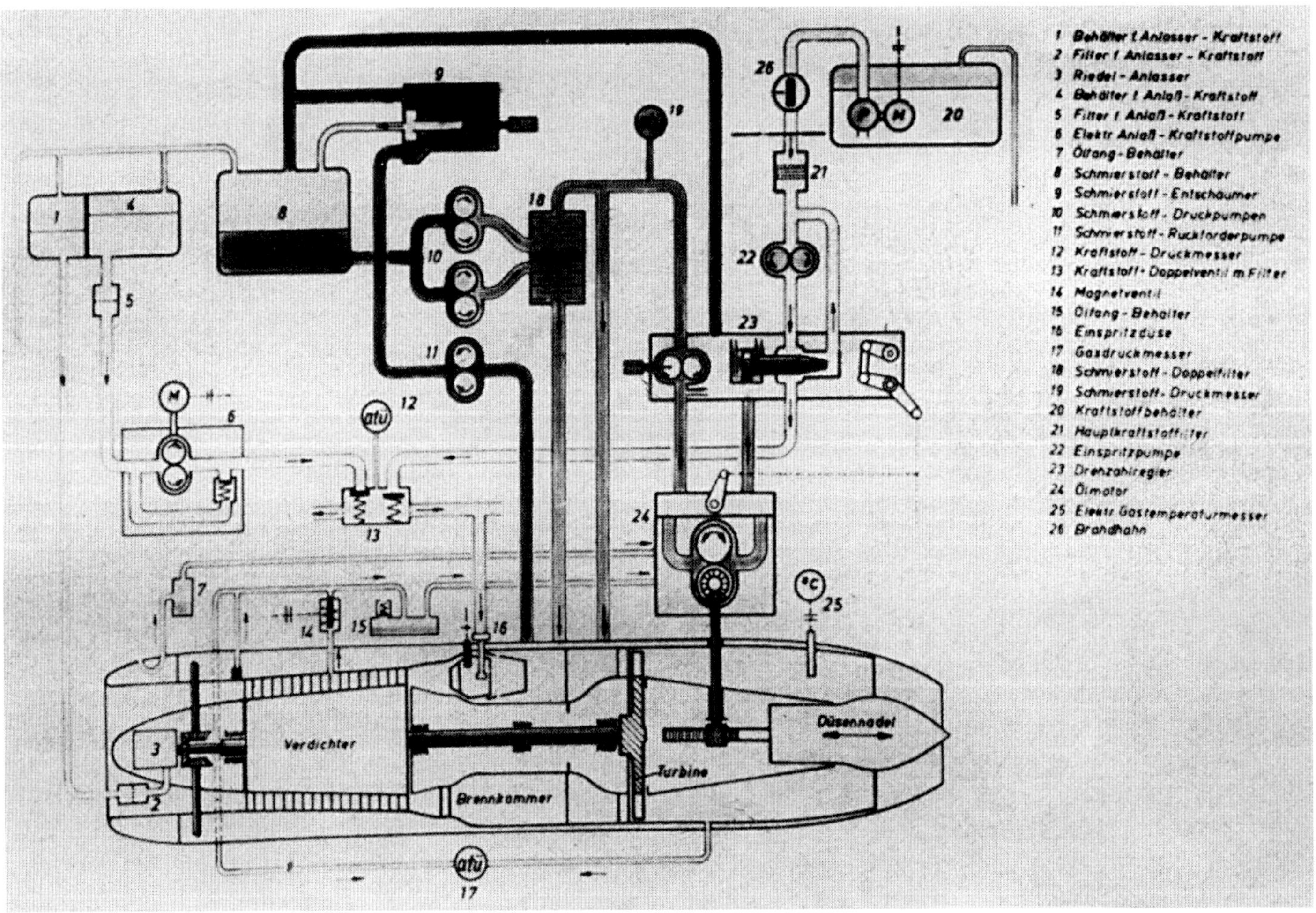

Operations diagram for the Junkers Jumo 109-004 B turbine. The Riedel starter, normally quite visible from the outside, is seen below number 3.

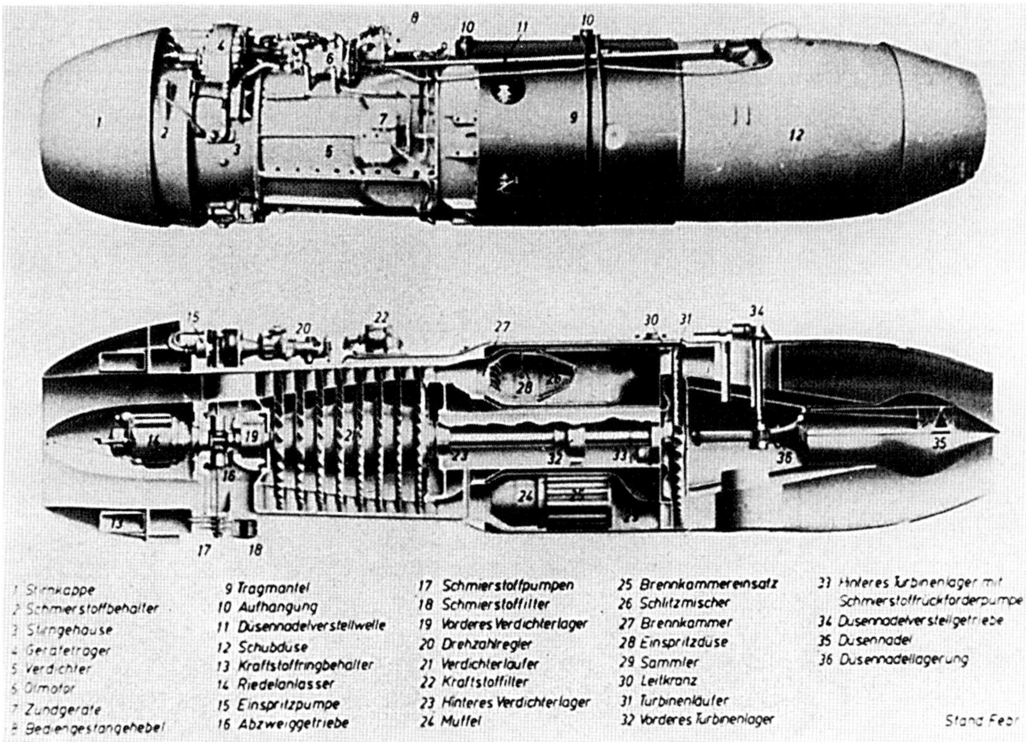

Junkers Jumo 109-004B, one showing the engine with the cover panels removed, the other showing a cutaway.

to overtake the BMW development in such a manner? For BMW's part, two experimental engines had been optimistically installed in the Messerschmitt Me 262 V1 in March of 1942. The first experimental takeoff under turbine power on 25 March 1942 was nearly a catastrophe when the two jet engines suddenly lost thrust during takeoff, then gave up the ghost altogether and simply froze up.

Thanks to the skills and abilities of the test pilot, Fritz Wendel, and the fact that the center engine (a Jumo 210 G piston engine) was also running as a safety measure, the test flight was able to be completed without incurring any damage. This disaster was a major setback for the BMW development program, so that it wasn't until the end of 1943 that a plane, an Arado Ar 234, could be successfully flown with a purely turbine engine. By this point in time the Junkers turbine was already in full-scale production.

BMW was never able to make up for lost ground after this, plus the Junkers turbine proved to be more reliable and robust. Since Junkers turbines were more powerful as well, these were favored for installation in the Me 262 standard fighter. On the other hand, when the BMW turbine was operationally used as an alternative, it was often used in pairs as a dual configuration, or in threes as a triple combination. Which brings us back to the Ju 287, for this was the first time that a triple combination had been planned for installation. This triple pack consisted of three BMW 109-003 A-1 engines and ran successfully on the test bench at the BMW Works in early 1945.

Junkers Ju 287 Rocket Assisted Takeoff Propulsion: HWK 109-501, HWK 109-502 and WASAG 109-522

So called rocket assisted takeoff was utilized in order to enable the Ju 287 to lift itself off the runway in the shortest time possible in the shortest distance possible. Due to their external shape, the troops sometimes called these RATO pods "*Kraftei*", or powered eggs (an appellation also given to the Me 163, by the way). Beneath the shell, however, there lurked a propulsion unit designed by the Walter Works in Kiel, based on a liquid fuel rocket engine using three different fuel components: T Z and Br+B (*T-Stoff* = hydrogen peroxide, *Z-Stoff* = liquid solution of natrium or potassium permanganate, *Br-Stoff* = gasoline, produced from benzine, and *B-Stoff* = hydrazine hydrate, acts as a catalyzer for the hydrogen peroxide).

These RATO pods were reusable and universally applicable since they could be jettisoned after running out of fuel and recovered by parachute. The pods, designated HWK 109-501, provided additional thrust in the amount of 1,000 kg with a burn time of 42 seconds or 1,500 kg thrust with a burn time of 30 seconds.

In order to make use of these RATO pods, the test bed had the HWK 109-501 hung beneath the engines. For the subsequent production version the initial plans called for the same configuration with the more powerful HWK 109-502 D, which was to have had a thrust output of 1,500 kg with a 42 second burn time. The drawing on the right shows the operation of the RATO pod.

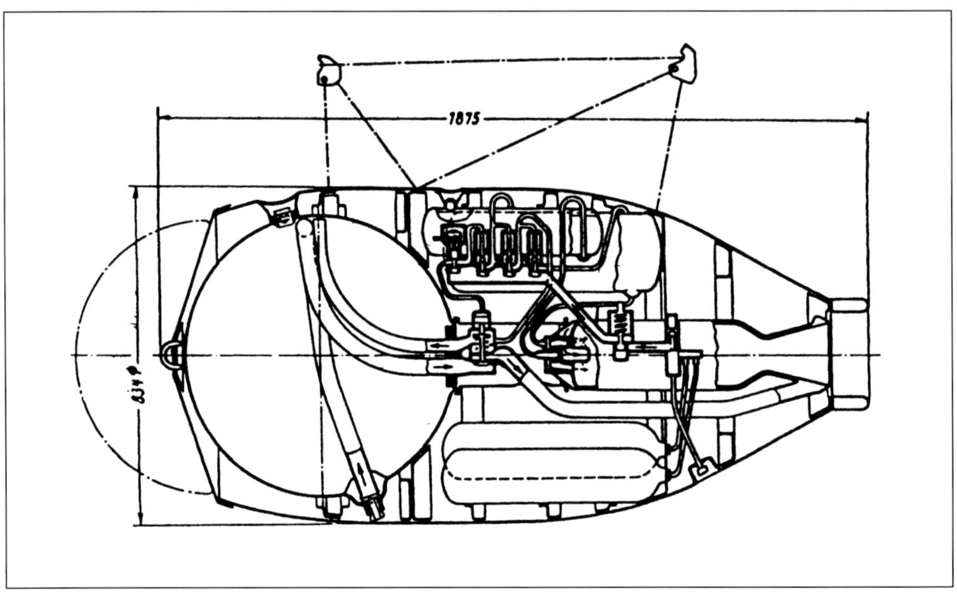

Overview of the Walter HWK 109-501 rocket-assisted takeoff pod, as was used in clusters of three on the Junkers Ju 287.

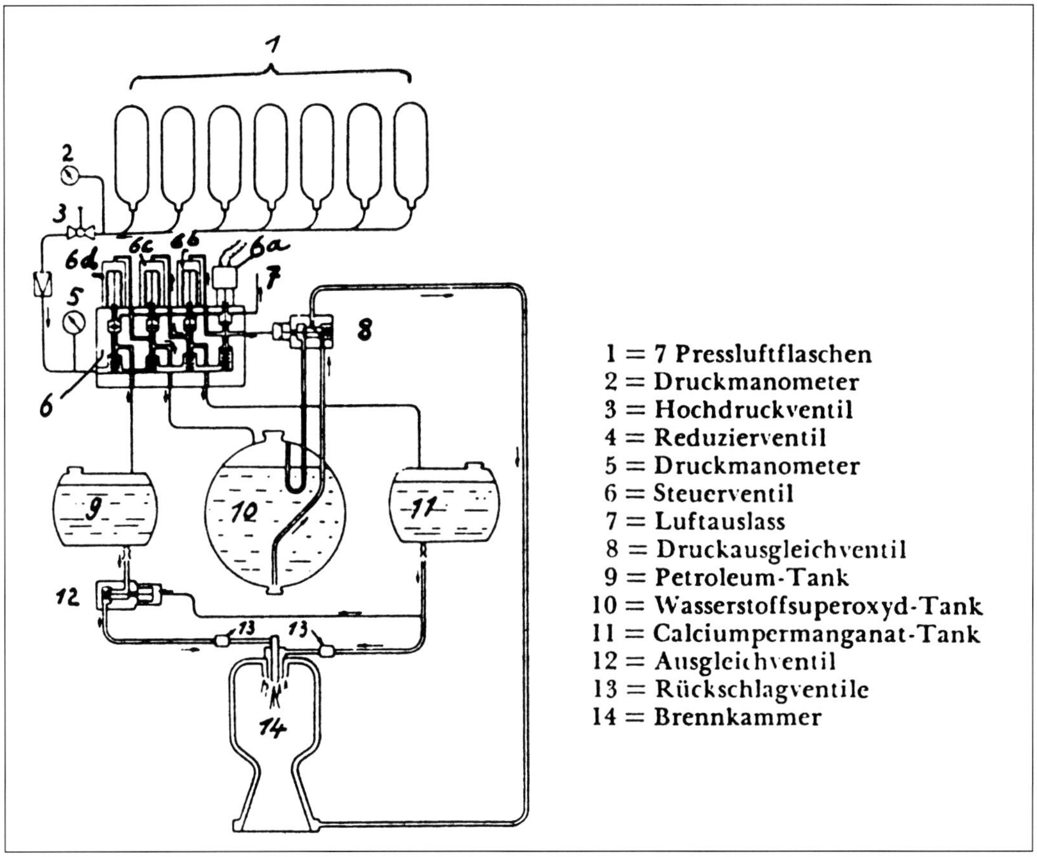

Control flow diagram of the Walter HWK 109-501 rocket pod. As can be seen, a full complement of compressed air bottles was needed to force the fuel through the pressure valves.

Key:
1. compressed air cylinders
2. pressure manometer
3. high pressure valve
4. reduction valve
5. pressure manometer
6. control valve
7. air vent
8. pressure equalization valve
9. petroleum tank
10. hydrogen peroxide tank
11. potassium permanganate tank
12. equalization valve
13. backflow valve
14. combustion chamber

Assembly sketch for the planned HWK 109-501 D RATO motor, a higher performance version with a thrust of 1,500 kg over 42 seconds.

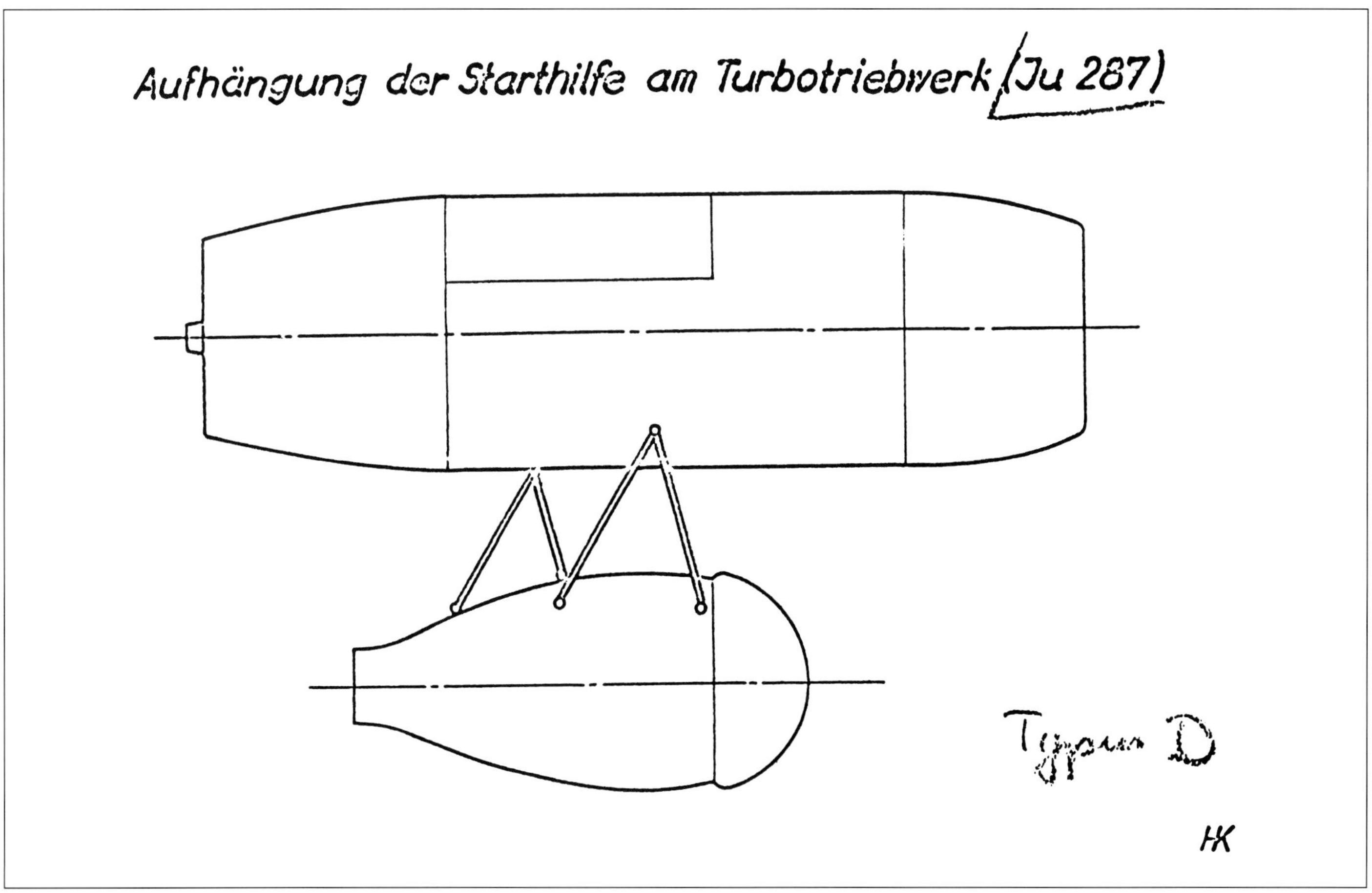

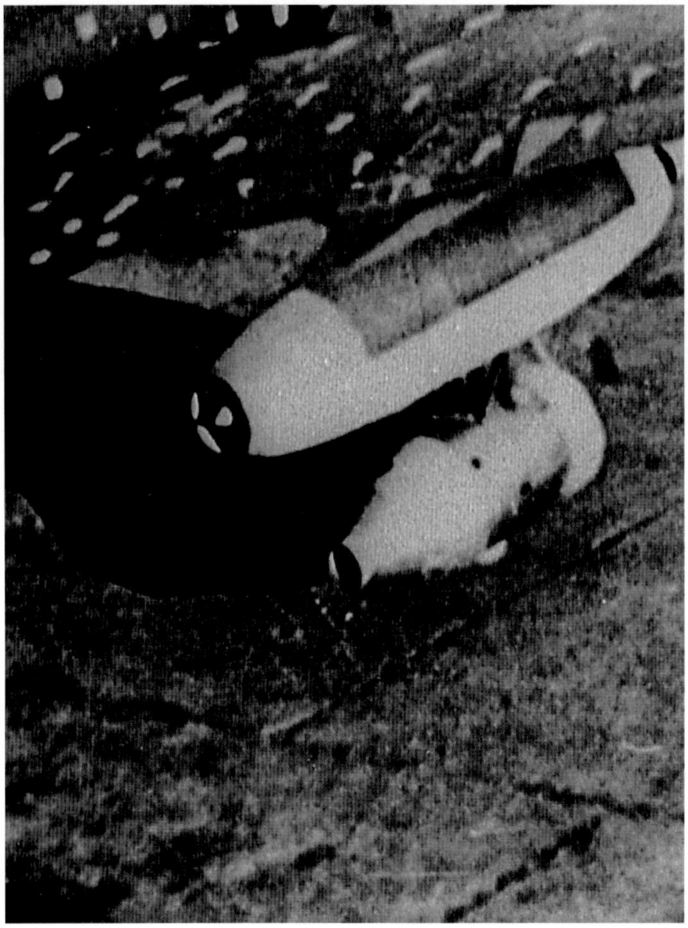

Detail photo of a Walter HWK 109-501 RATO pod beneath the starboard wing engine on the Ju 287 V1.

Shortly before the war's end, as it became more difficult to procure sufficient quantities of liquid fuel for the RATO motors, there began a search for a viable alternative. It was believed that three to seven solid fuel rockets built by the WASAG (Westfälische-Anhaltsche Sprengstoff AG) with the designation 109-522 would be entirely adequate for assisted takeoff when mounted to the tail of the Ju 287. These RATO units were powered by solid fuel consisting of diglycol and delivered 12 seconds of thrust at 1,000 kg. When the rockets burned out they were simply jettisoned. There was no intention to recycle them. Was this perhaps the beginning of the disposable society?

All kidding aside, just before the war ended there was no longer any chance of maintaining a recovery and collection service for the multitude of RATO packs dropped during massed launches of a bomber unit. Another criteria was probably the fact that the dimensions of the WASAG units were very small so that they could be bundled together on the Ju 287's tail. Although some thrust would be lost if one of them should malfunction, there would have been no asymmetrical increase in thrust causing an inadvertent yawing movement, such as would have been the case with the Walter units suspended beneath the engines.

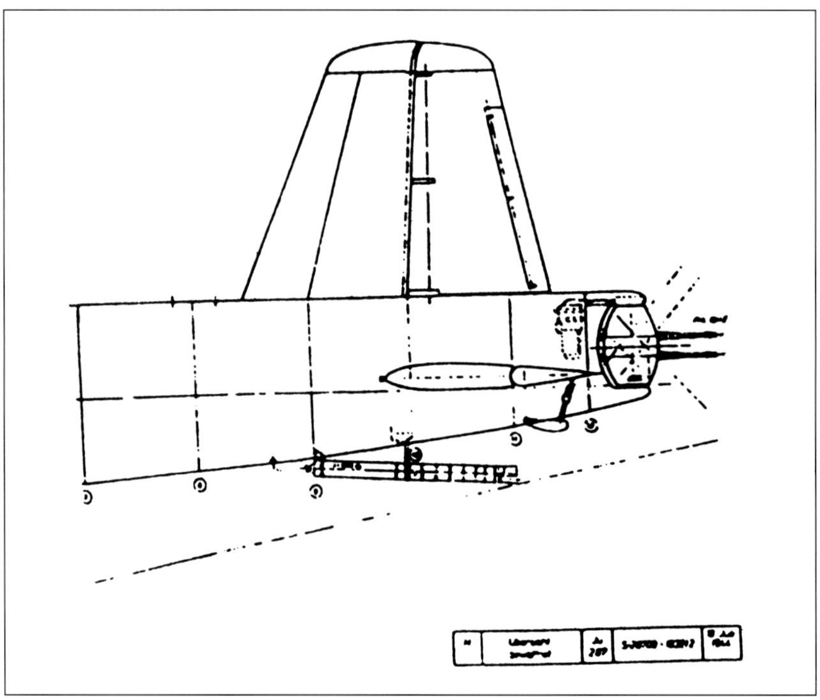

The WASAG solid fuel RATO booster, under consideration even as the war was drawing to a close, seen attached to the aft end of the Ju 287.

The Junkers Ju 287 (EF 131) in Model Kit Form

This chapter is for model builders and will take a look at the options for building the Ju 287 V1 - the simplest of the bunch, as well as the Ju 287 V2 - a little bit more difficult, and even the Ju 287 V3 - which poses the greatest challenge, into unique models in 1/72nd scale.

Something that might get the builder's heart beating a little faster is the knowledge that HUMA, which has once again established itself in the marketplace, released an injection molded model kit of the Ju 287 V1 in 1/72nd scale a few years ago. The finely detailed injection parts, which by the way is a specialty of HUMA, offers the possibility of an unusually detailed model. The detail of the parts compare quite favorably to the generally more expensive photo-etched parts and have the added advantage of being in plastic.

Nevertheless, there are some galling features that somewhat dampen the initial euphoria. First up is the horizontal stabilizer, which is set too low and must be raised by 3 mm in order to mirror the original. If this is bothersome, the elevator can be raised by puttying the original openings in the fuselage closed, cutting off the inserts on the elevators, and simply glueing them directly to the fuselage. Another defect is the not entirely accurate reproduction of the rudder. The differences from the full-size version are so minor, however, that it's not worth the effort to rework this piece. Much more serious, though, is the fact that the aft fuselage in no way reflects that of the original. The step in the fuselage caused by mating the empennage of the Ju 188 G-2 has not been reproduced and could only be rectified by reworking it using balsa wood, for

On the right is the model of the Ju 287 V1 from HUMA in 1/72nd scale, while next to it is the somewhat elderly Ju 287 V1 vacuum-formed kit from Intermodell bedecked with simulated wool tufts.

This Ju 287 V2, built from the HUMA kit with extra parts added, is an unusual addition to the popular modeling scale of 1/72nd that, with a little extra effort, can send many a collector's heart racing.

example. Every modeler must decide for himself whether he wants to correct this deficiency or will simply overlook it.

Back to the kit of the Ju 287 V1 itself. Overlooking the glaring errors in the kit, the model builder is offered a detailed kit with easy-to-follow instructions. The parts are generally flash-free, the fit is quite good, and there is virtually no need for extra work. One very important thing must be done before the fuselage is assembled: a generous amount of lead needs to be put into the forward fuselage area to weigh it down, otherwise the model will inevitably rest heavily on its tail. Another minor detail should also be mentioned. If the modeler desires to attach RATO pods to his model, he needs to make sure and attach only three; there is no known photo showing the Ju 287 V1 with four RATO engines. The paint scheme for the three Ju 287 V versions will be discussed at the end of this chapter.

Those who wish to make their lives more difficult can also opt for a vacuum-formed version of the Ju 287 V1. This, however, is only recommended for the experienced modeler who has several

Resin model of the Ju 287 V3 from Planet, built with much sweat and frayed nerves, shows an acceptable result after some modification work.

The following series of photographs of the Ju 287 built by Günter Sengfelder reveals just how true-to-life a modeler can make his model appear. The more realistic the background, the more natural looking can be the result. This first photo shows the Ju 287 in 1/15th scale taken from the side.

vacuum-formed models under his belt already. This kit has the correct fuselage shape, and it might be possible to combine the two kits into one.

If the intent is to create a Ju 287 V2 in miniature, the modeler should use the HUMA kit as the basis and drop the elevators 2 mm lower as mentioned above. With two exceptions, the model can then be built according to instructions. The first modification is to dispense with the tailwheel. The existing part is filed down and the hole in the fuselage is puttied and sanded. The second is that the engines are not mounted underneath the wings. Fortunate souls will have two extra BMW 109-003 twin engines left over from an Arado Ar 234 kit. The Matchbox kit of this type provides the option of an Arado Ar 234 B kit with two Jumo 109-004s or an Ar 234 C with two BMW 109-003 twin jets, leaving one engine set as a spare. Those who don't have these parts will have to bite the bullet and buy the kit if building the Ju 287 V2 is an absolute must, with the added bonus of sorts of having an Ar 234 B as leftover. Once the modeler has the BMW twin jets, these should be attached beneath the wings right next to the outer strut braces on the undercarriage, making sure that they're aligned parallel to the fuselage. The original openings for the engines should then be filled in with putty and sanded smooth. Putty will need to be used to blend the back of the engine pods into the trailing edge of the wings. This requires some patience, as the putty will need to be applied rather thickly and will take time to harden before it can be sanded. Don't forget that the supports for the main landing gear need to be turned inward. Once this work has been completed, a good paint job will give the builder a unique piece of German aviation history that can't be found in every collection. With a little more effort, a model of the triple-engine version can be built as well, although this will require considerable skill in mating the Junkers twins with the BMW single engine into a triple combination. But for those who are up to the challenge, best wishes on your venture. A modeler who really wants to go out of his way is fully justified in building the engineless Ju 287 V2 as it was left at the edge of the Brandis airfield, with only some putty work necessary to close up the unneeded openings. Don't forget the engine mounts (see picture on page 73) beneath the wings for the BMW twins, which can be made from scrap plastic.

For the Ju 287 V3, there's a Planet resin kit on the market that also permits building the EF 131 V1. To be sure, this isn't in everyone's taste, but it is the only kit of the Ju 287 V3 in 1/72nd scale. If the interested modeler is initially put off by the kit's defects and the idea of working in resin, the high cost of the kit, easily

Another photo of the model taken from a front angle offers an even more realistic appearance.

at 50 Euros, will only add to any reservations. But there is little alternative for those who don't have a lot of accessories in their toolbox and absolutely must have a Ju 287 V3 or an EF 131 in their collection. Not only that, but the model builder must have particularly strong nerves if he wants to put together the two vacuum-formed halves of the canopy found in this kit without any lasting damage to body and soul. A person who, despite these drawbacks, still dares to tackle this kit - which really is recommended only for the most experienced modeler - will have as a result not only a truly important model, but one that is certainly unique in his collection. During construction, attention needs to be given to the forward fuselage area, where a generous amount of lead weight must be included in order to prevent the model from becoming a tail dragger instead of resting properly on its tricycle undercarriage. Additionally, the engine configuration of the model doesn't accurately reflect that of the prototype. However, this defect can be easily corrected by attaching the single engine under the twin engines, thus obtaining the desired triple combination. Don't forget to file off the fuselage points for the single engines, putty the holes in the forward fuselage and then sand the area smooth. All in all, it should be mentioned that there's a lot of putty work involved in this kit.

The cunning modeler could obtain a vacuum kit of the Ju 287 V1 as well as another vacuum kit of the Ju 388 K, take the wings from the Ju 287 V1, the cockpit and fuselage together with rudder from the Ju 388 K, and build the elevators from vacuum-formed parts and the triple- engine combinations using parts from the spares box. Again, the latter is accomplished by taking two leftover twin engine combos and glueing/puttying a single engine unit to them. The individual pieces are aligned with each other using available three-view drawings, putty is applied, and then the whole thing can be superdetailed using parts from the spares box. In this manner - obviously not recommended for everyone - a modeler can, with a lot of work and time invested, obtain a presentable model that is easily on par with the much more expensive resin model kit alternative.

Painting the Ju 287 is a rather tricky affair, particularly since the meager two color scheme in HUMA's box has little to do with reality. Photos are a must in this case, and a detailed study shows that the Ju 287 V1 most likely was painted as follows:

At first glance, it's impossible to differentiate this picture from the real thing. Masterfully placed in its setting by Günter Sengfelder, the 'real' photo effect is enhanced by the use of black-and-white film.

Upper surfaces: two-tone green using RLM 70 and RLM 71, with gray patches on the upper rudder in RLM 02.

Lower surfaces: there is considerable speculation here - either *Hellblau* RLM 65 or RLM 76 or *Grau* RLM 02, with the latter being the most probable color.

Other: side of the fuselage between the fuselage and undercarriage - black protective paint due to the hot exhaust from the fuselage-mounted engines.

Starboard fuselage - checkerboard pattern in *Rot* RLM 23 and *Alu* RLM 01 or *Schwarz* RLM 22 and *Alu* RLM 01, with the latter once again probably being the most accurate as it has been verified by eyewitnesses.

The cover plates for the engines changed, as can be seen in different photographs, so the modeler is spoiled for choice and will have to decide on one of the several variations. The dark areas were undoubtedly *Grün* RLM 71 and RLM 72 and the lighter areas were in *Grau* RLM 02.

The undercarriage spats for the main wheels as well as that of the nose gear and the "trouser legs" on the Ju 287 V1 were painted in *Grün* RLM 71. Alternatively, the trouser legs on the Ju 287 V2 were in *Grau* RLM 02. Other colors were probably similar to that of the Ju 287 V1.

The camouflage markings of the Ju 287 V3, if it would have been completed, is purely speculative in nature, so that a modeler can let his fantasies run wild here. However, at this time, the standard camouflage scheme was being applied regularly. This consisted of upper surfaces in *Dunkelgrün* RLM 81, *Hellgrün* RLM 82, and lower surfaces in *Hellblau* RLM 65 (from 1941 onward *Hellblau* RLM 65 was approximately the same as *Grau* RLM 02).

As we can see, it is possible for a person to turn the colors of the Ju 287 into a science unto itself, and whether or not his concern for such detail is accurate can only be determined by a color photo or a paint chip. Certainly, as the war was drawing to a close there was not much value placed on proper camouflage colors; whatever was on hand was used. Because of all the excitement generated by various three-views, a larger scale model can also be built without difficulty. One man who's already attempted this with much success is Günter Sengfelder, who has created quite a beautiful model in 1/15th scale (see photos on pages 120 and 121).

The Ju 287's Twin Sister
The EF 131 Developmental Airplane

When the Russians took over control as the new masters in Dessau and other Junkers works on 1 July 1945, they showed keen interest in the developmental work on the Junkers Ju 287 as well as development of the Jumo 109-004 turbojet engine. All the significant personnel were soon found and rounded up. Several developmental groups were formed and organized according to whether they were working on the airframe or the propulsion technology.

Since in the case of the Ju 287 the planning and production documentation was only fragmentary, the collected designers and scientists were to design a new aircraft on the basis of the unfinished Ju 287 that would carry the designation EF 131 (EF = Entwicklungsflugzeug, developmental aircraft).

In order to carry out Operation EF 131 it was first necessary to bring the wind tunnels, production facilities, etc. back into functioning order. The next step was to subordinate the design team under the protection of OKB 1, or Design Bureau 1 (experimental design bureau). At this point it's worth taking a little look at the roots, purpose and use of the OKBs.

When the Russian troops first crossed into German territory at the beginning of 1945 they were justifiably hopeful of soon winning the war. At this time they began seriously thinking of how best to use the scientific potential of the Germans for their own purposes after the war was won. Similar to the Americans' Operation Paperclip, the Russians formed a commission for seeking out German scientists as well as for exploring the production facilities of the aviation industry in the areas they occupied. This spawned two camps with very differing views on how to carry out their missions. The one group was of the opinion that the proper way to proceed would be to dismantle the production facilities in Germany and ship them to the Soviet Union, with the associated scientists and designers brought along at the same time to keep things simple. The other group, on the other hand, held the opinion that both personnel as well as equipment should remain on site and that production should resume on German soil under the supervision of the Russians, who could monitor future developments and use the results for the benefit of their own country. It was agreed that any approach would be based on importance or, in the case of the EF 131, to use both approaches. The personnel were initially left alone in their familiar surroundings in order to then consider themselves as guests of the Soviet Union over the positive course of development and then ship off men and material to Russia without further ado.

But let's return back to the EF 131 where, in the meantime, the responsibility for the program as well as overall management of the EF 131's design work was given to *Dipl.-Ing.* Brunolf Baade, with *Dipl.-Ing.* Fritz Freytag as his deputy working on design refinements and *Dipl.-Ing.* Hans Wocke. This team, fleshed out by even more former Junkers employees, began working on the design of the Project EF 131 as early as September 1945 and soon had made rapid, visible strides in the program, as we shall see in the next chapter.

This fictional air-to-air drawing of the EF 131 by Uwe Jack once graced the cover of Aviatik magazine, volume 2.

Devlopment of the EF 131

After an initial attempt to reconstruct the Ju 287 from any remaining documentary fragments on the Ju 287 series, the people in Dessau decided to build a new machine on the basis of the type that would be quite close to the external appearance of the Ju 287. This development, now called the EF 131, quickly took shape thanks to the existing documents. It was decided to initially build only three prototypes, but only the EF 131 V1 and the EF 131 V3 would be made available for flight trials. The EF 131 V2, on the other hand, would be used for static and dynamic stress testing only. It soon became obvious that from a purely external perspective the EF 131 was virtually indistinguishable form the Ju 287 V3 and shared other common traits in its layout, so it is entirely possible that the remaining parts of the Ju 287 V3 were finished/added to in order to use this airframe for stress tests.

In July 1946 a mockup in 1:1 scale was completed and displayed for viewing. A commission under the direction of Lt. Gen.

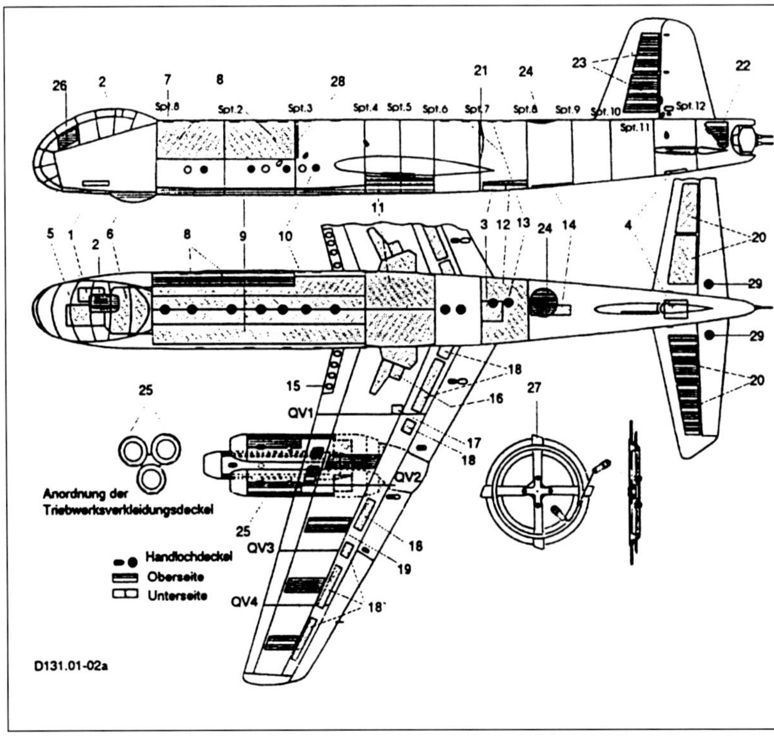

Key:
1. forward entry hatch
2. forward exit hatch
3. entry hatch forward of bulkhead 7
4. parabrake cover and entry hatch forward of bulkhead 11
5. strut door (nose gear)
6. nose gear doors
7. fuel point cover
8. compartment access cover
9. bomb bay doors
10. bomb rack inspection port cover
11. undercarriage doors
12. compartment access cover
13. compartment inspection port cover
14. fuel dump access cover
15. leading edge slats maintenance access covers
16. strut door (main undercarriage)
17. landing light access cover
18. control surface access hatches
19. wing access covers
20. elevator access covers
21. bulkhead 7 compartment hatch
22. ammunition access hatch
23. vertical stabilizer access covers
24. radio equipment access cover (DF)
25. engine nacelle access covers
26. opening window
27. snap closure
28. removable upper bulkhead
29. elevator inspection cover

Overview showing the covers and hatches from the service manual for the EF 131, by Dr. Peter Korell.

V. A. Ushakov inspected the mockup and gave the green light for construction of the EF 131. Without waiting for the commission's decision, however, work had already begun on the EF 131 V1's construction in January of 1946. In his book "*Deutsche Spuren in der sowjetischen Luftfahrtgeschichte*" (German Traces in Soviet Aviation History), Dmitri Alexeevich Sobolev writes on page 153 that the wings of the Ju 287 V2 were reputedly used on the EF 131. But since Mr. Sobolev often interchanged the Ju 287 V2 with the V3 and apparently had no idea that the Ju 287 V2 was used as a test bed in his well documented book, it seems as though a few things became confused here or were not organized properly. What is closer to reality is probably that the Ju 287 V3 was used as a stress test airframe and the undamaged wings of the Ju 287 V2 were indeed used for the Ju 287 V3 in those stress tests.

For its powerplant, the EF 131 was to have been equipped with two triple combination units consisting of three Junkers Jumo 109-004 B each, giving it a speed of 860 km/h. One more thing needs to be said about the engines: these were designed by a different developmental group from the former Junkers Flugmotorenfabrik. Like the Ju 287, these were reconstructed and later produced in a small run of 30 units in Russia. One member of this team was *Dipl.-Ing.* Heinrich Theuer, who is still alive today. As Mr. Theuer reports, an engine test bench was set up alongside the assembly area for the EF 131, where the triple combination powerplants were tested in several engine runups. In doing so, it was discovered that, in spite of all the fears, the three bundled engines hardly affected each other negatively. After all, unlike the BMW triples, there had been no previous attempt to cluster three Junkers engines together. The

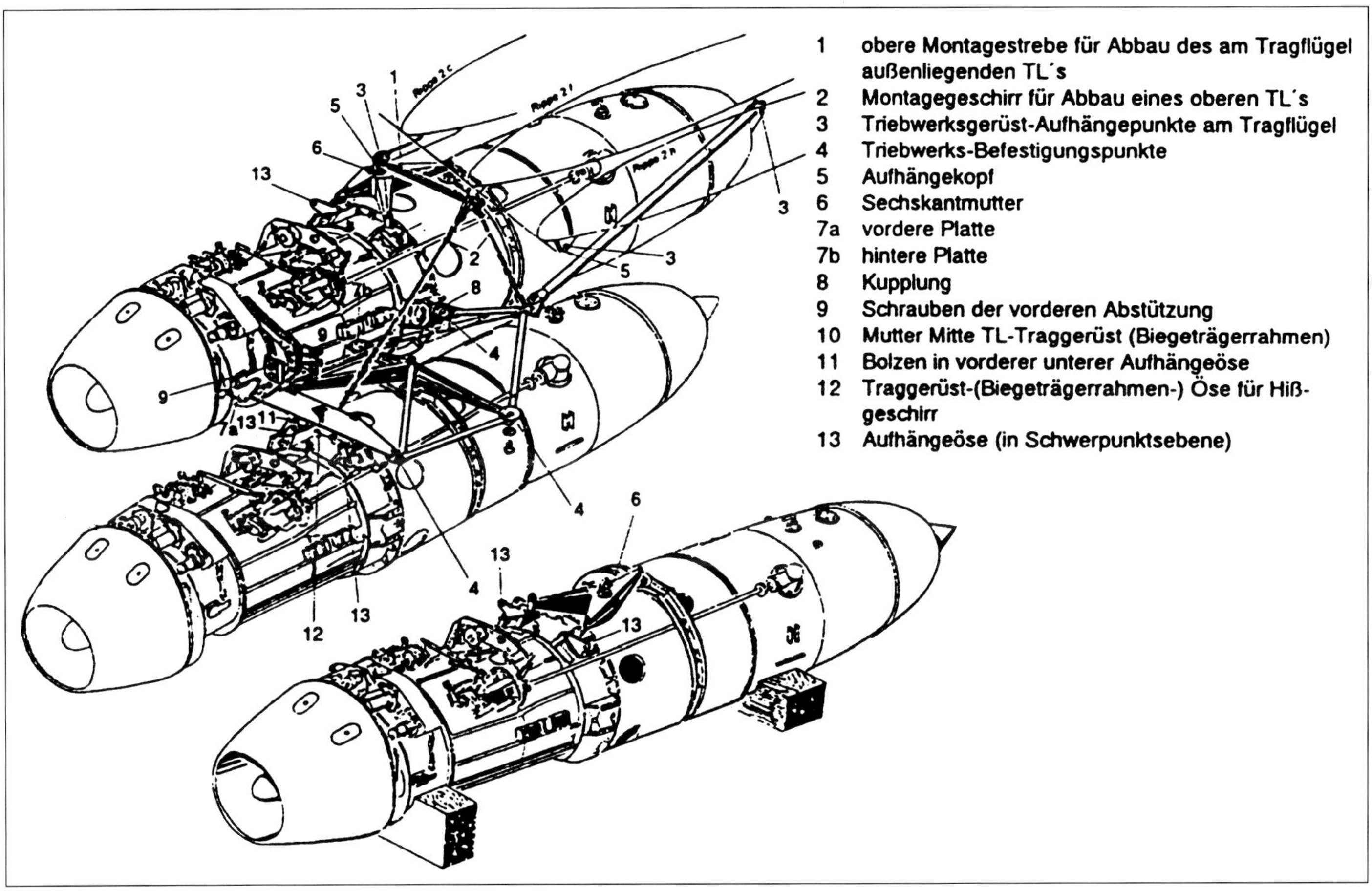

Drawing of the engine framework with mounting hardware for the triple engine configuration, consisting of three Jumo 109-004 B turbines (excerpt from the EF 131 service manual by Dr. Peter Korell).

Key:

1. upper assembly brace for removing turbojet from outer wing point
2. assembly framework for removing upper turbojet
3. engine frame to wing attachment point
4. engine anchor point
5. suspension lug
6. hexagonal nut
7a. forward plate
7b. aft plate
8. coupling
9. forward support bolts
10. center engine frame nut (flex support frame)
11. forward lower eyelet attachment bolts
12. support frame (flex) eyelet for hoist equipment
13. attachment eyelet (at center of gravity point)

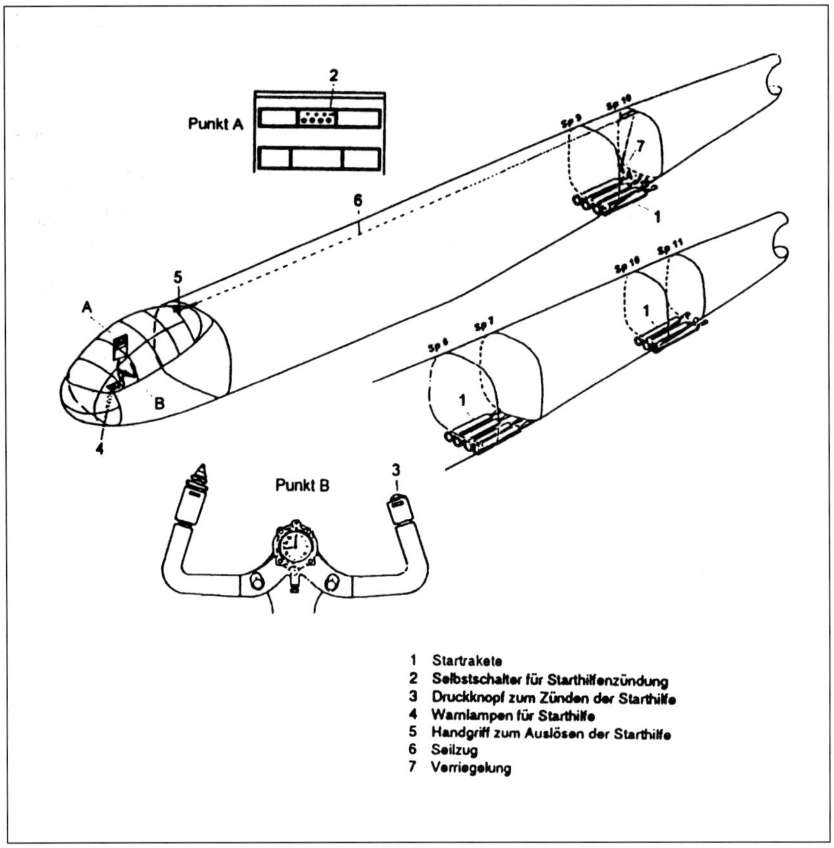

Sketch showing the attachment of the RATO packs on the EF 131's fuselage (excerpt from the EF 131 service manual by Dr. Peter Korell).
Key:
1. booster rockets
2. automatic switches for booster rocket ignition
3. booster rocket ignition button
4. booster rocket warning light
5. booster rocket release handle
6. cable
7. locking mechanism

Sketch showing the braking parachute, deployed to shorten the roll-out on landing (excerpt from the EF 131 service manual by Dr. Peter Korell).
Key:
1. handle
2. cable
3. parabrake
4. retainer cable
5. retainer cable link coupling
6. retainer cable separator coupling
7. parabrake compartment
8. parabrake compartment cover
9. retaining pin

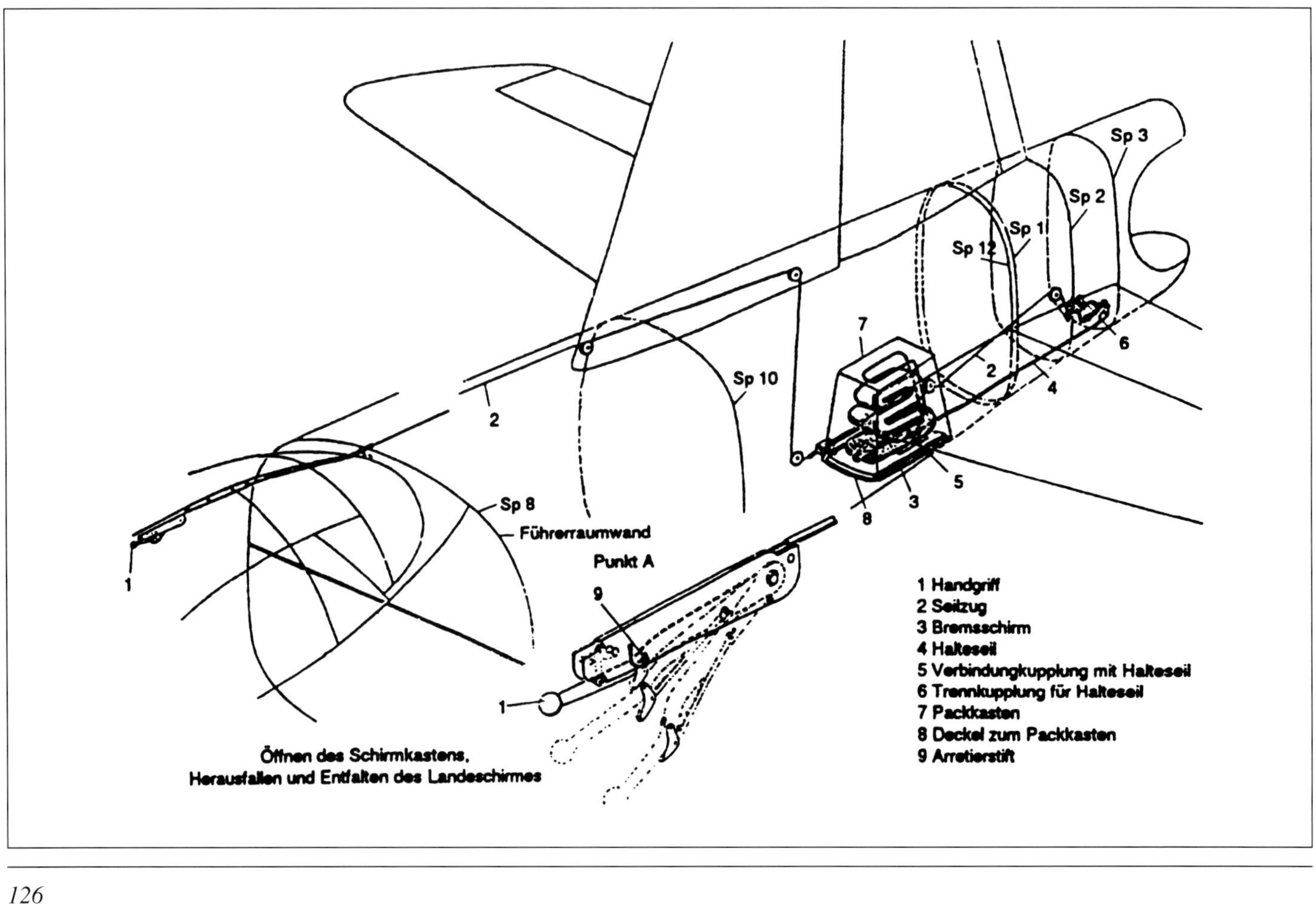

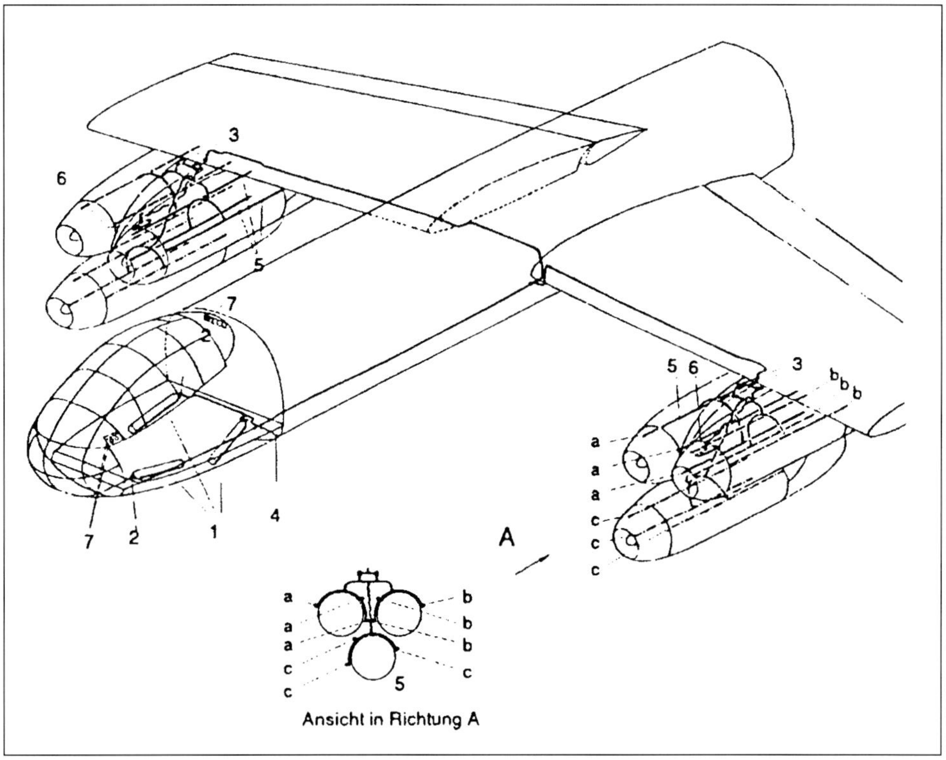

Sketch showing the fire extinguisher system for the EF 131's engines (excerpt from the EF 131 service manual by Dr. Peter Korell).
Key:
1. steel alloy cylinders
2. cylinder valves
3. extinguisher vents
4. excess pressure valve
5. engine spray tube
6. air vent actuation cylinder
7. fire warning indicator panel

combined thrust at 2,600 kg was, however, somewhat less than the total of the three single outputs at 900 kg each. Heinrich Theuer was one of the engine specialists who later went to the Soviet Union for many years, but more of this later.

With a bomb payload of 2,000 kg a range of 1,050 km could safely be reached. The three-man crew sat in a spacious, fully-glazed cockpit offering excellent all-round visibility. The wings borrowed from the Ju 287 had a negative sweep of 19%50". Furthermore, the wings were fitted with automatic leading edge slats and split flaps on the trailing edge, with the slats being of a new design compared to those on the Ju 287. They had an automatic extension feature that would pull the aircraft out of a critical angle of attack.. The empennage was designed in such a manner that the elevators were positioned in the exhaust flow from the jet engines, obviating the need for a de-icer system for the elevators. Compared to the Ju 287, the fuselage was some 2.5 m longer. Furthermore, the EF 131 was to have an automatic gun barbette in the tail, fitted with two machine guns similar to the Ju 287. In addition, the extremely effective fire extinguisher system from the Ju 287 was utilized (see drawing above). Unlike the Ju 287, though, four or seven RATO pods were capable of being fitted under the fuselage (see drawing on left), each with 1,000 kg thrust - something that would shorten the takeoff run considerably. At the opposite end of the flight profile, the brake 'chute of the Ju 287 was also borrowed to shorten the landing roll-out (see drawing on left). So much for the brief description of the EF 131. It is pointless to discuss the development down to the finest detail, for it would only be a repetition of the Ju 287's development under different conditions; therefore, we can permit ourselves to limit the discussion only to the most critical areas.

On 12 August 1946 the EF 131's assembly was complete, and it was ready for ground trials. Following a handful of taxi trials in Dessau and after the leveling adjustments had been made, the EF 131 was turned over to flight testing on 16 August 1946, but it never made it. To ensure secrecy, the test flights were not to be carried out in front of the Western Allies' noses. It was decided to disassemble the machine and, together with all necessary materials - to most likely include the Ju 287 V3 as well - and all necessary personnel, move everything to the flight test center at Podberezhye, about 100 km north of Moscow. Almost as an afterthought, the production facilities were also disassembled and carted off to the Russian hinterland. The nighttime shipment took place on 16 September 1946, using 19 railroad cars pulled by military train no. 300/1003, with the destination of the Lyotno Ispitatelni Institut, the Soviet flight test center. Also in this shipment was the EF 126 experimental plane, a single jet strike aircraft with a Schmidt-Argus ramjet mounted on top, similar to the V1.

The same fate awaited the engine specialists; they were lodged near the township of Kuybishev on the Volga. Since the capacity there was limited, some had to be sent to the factories at Tushino and Chimki. All this took place without any prior notice and without concern for the willingness of those involved. Heinrich Theuer reports on the rough treatment he was given during this forced deportation:

During the early morning hours of 22 October 1946 I was rudely awoken from my sleep at around 5 AM by a loud knocking. Without beating around the bush, several armed Russian soldiers escorting a few higher ranking officers and a translator, made it clear to me that I must immediately pack up my important things as I was being moved. Without even the slightest clue, a short time later I found myself together with several of my colleagues and other Junkers employees at the Dessau train station. With dark forebodings we found ourselves herded into sealed and guarded compartments. The worst rumors began circulating, and all of us feared for our lives. The seemingly endless journey lasted eight days and kept us heading eastwards. It grew increasingly colder and we were freezing terribly. But eventually this horror ride came to an end; we were shocked by the miserable dump we found ourselves in, which became our second home for many long years. We settled down as well as could be expected and tried to make the best of things.

Kuybishev was the name of the place where we were to begin our work. Each of us had a laughable 8 m^2 of living space, and the wind blew through the cracks in the windows. We filled those cracks with newspaper in September and didn't remove them until the spring - no ventilation the whole time!!! Despite this, with outside temperatures below -35% C our work areas were only +4% even with heating. We couldn't even think about working at such temperatures. Many of my colleagues took their lives under these horrible conditions by going out into the cold, taking off their warm fur hats and, as the fluids in their brains froze, dying a quick and silent death. I, too, was tempted to do the same, but again and again the thought that I would eventually return home kept me from doing so. Finally, in late 1953, with Stalin dead and buried, I received the welcome news that I could finally turn my back on this wretched place.

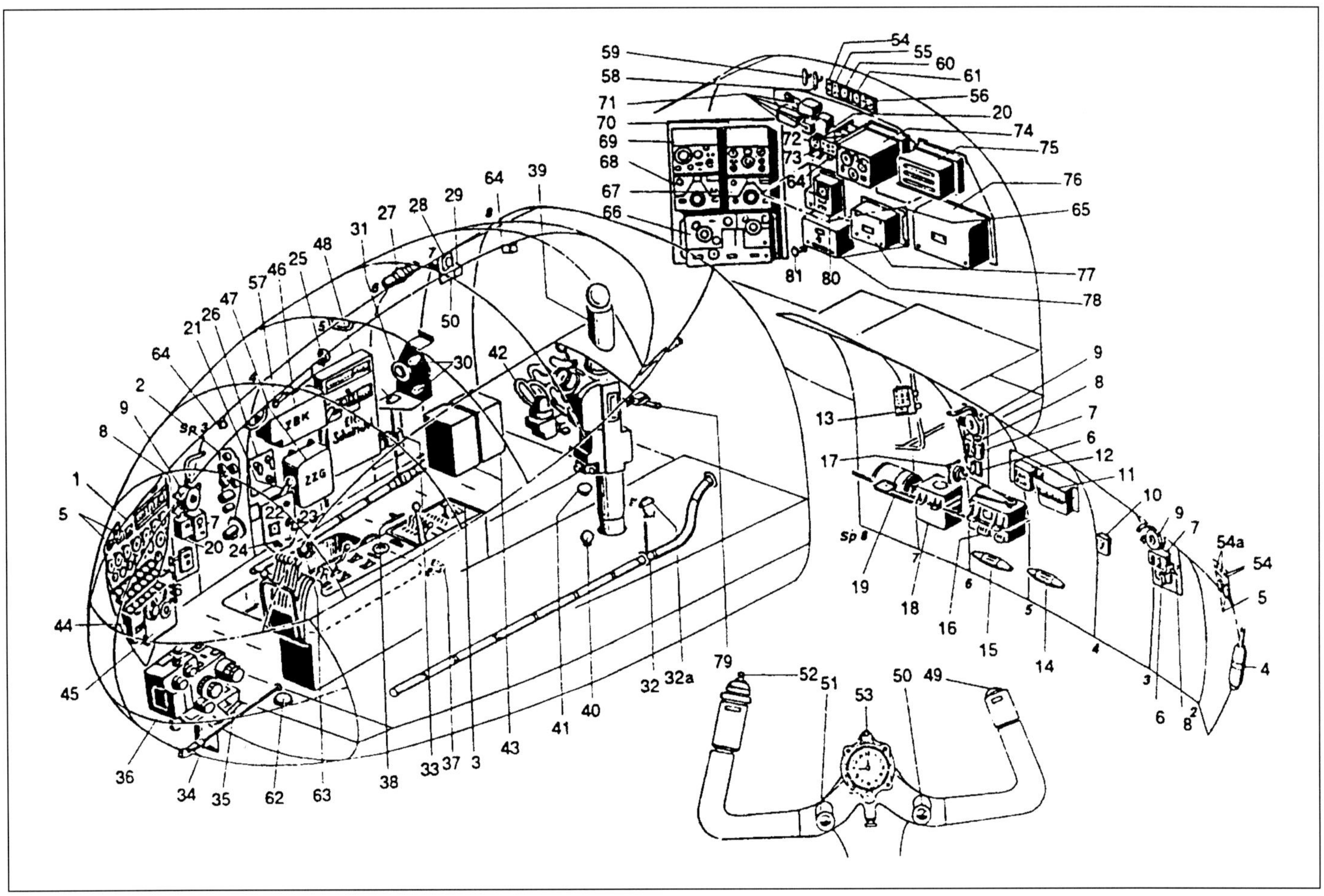

Sketch showing the equipment/instrumentation layout in the EF 131's crew compartment (excerpt from the EF 131 service manual by Dr. Peter Korell).

1. engine panel
2. IFR panel
3. instrument console
4. variometer leveling reservoir
5. oxygen pressure gauge and monitor
6. clothing heater connector and rheostat
7. radio headphone connection point (4DB 11)
8. oxygen system cutoff valve
9. oxygen mask
10. radio headset connecting jack (AD 17 Y)
11. map case
12. first aid kit
13. ready panel and ammunition counter for tail gun
14. Leonhard transformer for automatic pilot
15. AC-DC transformer for autopilot
16. control instrumentation
17. damping gyro regulator
18. directional gyroscope
19. master horizontal gyroscope
20. loop antenna control switch
21. emergency exit handle, activated by compressed air
22. compressed air gauge
23. compressed air cutoff valve
24. compressed air filler connector
25. hand lamp plug
26. 3 plugs: for ZZG connection
27. hand lamp
28. FuG 16 ZY long/short range selector switch
29. radioman microphone button
30. selector handle for emergency hydraulic pressure system
31. hydraulic oil pressure system depressurization valve
32. cockpit pressure cutoff valve
32a. cockpit airflow regulation lever
33. UV lighting rheostats
34. dust vent
35. automatic pilot direction indicator
36. Lotfe (*Lotfernrohr*) bomb sight
37. entry hatch emergency release handle
38. hand pump for emergency hydraulic system
39. periscope
40. protective cover removal handle
41. ventral/dorsal view switch knob
42. tail gun controller
43. accumulator for electrical on-board communications system
44. ZSK (*Zündschaltkasten*- ignition switch box)
45. RAB (*Reihenabwurfbediengerät* - bomb salvo control panel)
46. ZBK (*Zünderbatterienkasten* - ignition battery box)
47. ZZG (*Zünderzusatzgerät* - ignition adapter)
48. electronic instrument panel
49. RATO ignition button
50. bomb release button
51. microphone button
52. heading indicator switch
53. clock
54a. RATO warning light
55. fire extinguisher activation button
56. AFN 2 radio navigation instrument
57. parabrake deployment/jettison handle
58. RATO activation handle
59. fuel dump release handle
60. radioman oxygen gauge
61. oxygen monitor
62. OK 42 magnetic compass
63. vertical compass
64. UV light
65. map table
66. FuG 16 ZY
67. transmitter - long
68. transmitter - short
69. receiver - short }
70. homing receiver } FuG 10 P
71. 5 connector jacks
72. trailing antenna cap switch
73. SchK 17 switch box
74. FBG 3 control panel
75. tube compartment
76. U 10 S transformer
77. U 17 transformer
78. U 11 transformer
79. radio
80. homing converter
81. lever to fill the emergency hydraulic pressure network with fuel

Breakout of equipment and instrumentation in the EF 131's crew compartment.

It is difficult to return to the main theme of this chapter after this moving account by Mr. Theuer, who was responsible for jet engine thermodynamics. Hardly had men and material arrived before work began on building the EF 131 as well as the Ju 287 V3 stress test airframe. After that, though, testing had to be postponed yet again. Durability testing revealed that the fuselage was only at 52% of the structural norm - sabotage was suspected. Because of the reinforcements to the airframe that this necessitated the project was delayed by a further two months. Ultimately, it took until 23 May 1947 until the EF 131 was finally ready for its maiden flight.

Postwar Trials of the EF 131 in Russia

Since both Siegfried Holzbaur and Hans Joachim Pancherz, two of the most prominent test pilots for Junkers, managed to escape the Russian occupiers, pilot Paul Jülge was compelled to follow as their successor. So it wasn't surprising that he was also drafted for testing the EF 131. ON 2 May 1947 he found himself in the pilot's seat of the EF 131 alongside flight engineer Günter Schröter, who is still alive today, and mechanic Walter Moses. The following report was compiled from the experiences on the first flight:

> Taxiing out for takeoff, the flight of 15 minutes total duration, and the landing went smoothly. The aircraft was piloted by German pilot Paul Jülge - today a test pilot with Experimental Works No. 1 (author's note: unlike test flights in Germany, where the test pilots wrote the flight reports themselves, reports in Russia were drafted by Russian personnel. Experimental Works No. 1 refers to the complex outside Moscow not far from the flight test center at Podberesye. Based on the opinion of the pilot and also from impressions by ground observers, the aircraft exhibited good handling characteristics on this, its first flight. It lifted off at a speed of 250 km/h and a takeoff weight of 17 metric tons without any problems. The maximum speed allowed for the first flight was 350 km/h, with the landing speed at 220 km/h. Level flight was conducted at an altitude of 1,400 m. After landing there was a minor accident caused by a bolt on one of the struts to the port main undercarriage shearing off. As a consequence, the aircraft drifted to the left and scraped the ground with one of the three engines. The airplane's damaged engine was changed and the minor damage caused by the collapse of the gear was corrected. Flights will continue in three to four days.

This thoroughly positive report essentially describes only a single test flight consisting of no more than one airfield circuit and omits to mention that the cause of the collapsed undercarriage was an improperly constructed landing gear. For safety reasons, the landing gear remained extended throughout the flight. Any risks were to be avoided during this first flight until it could be determined how the airplane behaved in the air, plus the systems needed to be checked out on takeoff and landing. The ribbon parabrake must have played a significant role in the minimal damage caused in the accident, with the traction force prevented the aircraft from breaking off the runway and potentially worse damage being caused. In addition, the braking 'chute ensured a quicker deceleration during the roll out. After a few weeks the repairs were complete, although this took more time than conveyed in the report. By the 31st of July 1947 a second test flight lasting 30 minutes had been carried out.

In two minutes the aircraft climbed to an altitude of 1,200 meters. Level flight took place at speeds of 270, 300, and 320 km/h with landing gear in the extended position. The aircraft is easy to control and stabile in flight. No vibration was encountered in any flight attitude, nor was there any flutter noted. The pilot flew turns with a bank of 35% and on landing gave the wings a slip of three to four degrees. Here, too, the control was particularly good. Landing approach took place at a speed of 235 km/h as measured by the pitot tube. The airfield perimeter was overflown at a speed of 230 km/h. Touchdown occurred at 200 km/h. Engine acceleration was good. Two seconds after landing the pilot applied half brakes, and after 300 m the braking 'chute was deployed. No jerking was noted; all radio equipment functioned normally. The takeoff run was 1,200 meters, and the rollout stretch was 850 to 900 m. The pilot's general impression: the machine is stabile, easily controlled at speeds of up to 350 km/h and does not differ from production Junkers aircraft either in flight or in control handling.

This second flight says something a little more than just the general flight characteristics of the aircraft. What is both notable as well as unusual is the fact that, on the one hand a slip was made with a new, as yet untested aircraft, and on the other hand - even more unusually - it was even attempted at all in view of being such a large aircraft. Anyone who has ever held a pilot's license knows that it's advisable not to attempt a slip with every airplane on landing in order to lose speed and altitude, since the handling can differ widely from aircraft to aircraft. Particularly with faster aircraft, aerodynamic forces can come into play that can easily damage the airframe. Subsequent testing of the EF 131 indeed revealed minor problems such as oscillations in the nose gear and stabilizer vibrations. These defects proved to be quite troublesome and much time was given over to correcting them. This was certainly the reason behind the fact that up until October 1947 the EF 131 had only logged a meager 4.5 hrs of flight time. Although no periscope can be seen on the only known photograph of the EF 131, *Herr* Schröter claims that a periscope sight was indeed installed, and it was through this periscope that the bowing of the fuselage was measured during various flight maneuvers carried out on several experimental flights.

The arrival of winter brought further problems with it. Since the EF 131 did not have its own hangar, it was forced to camp out in the open under snowy and frosty conditions. This naturally had a negative effect on any additional testing operations. The most serious of the problems was the fact that the engines proved particularly unwilling to do their duty in frosty weather when the tem-

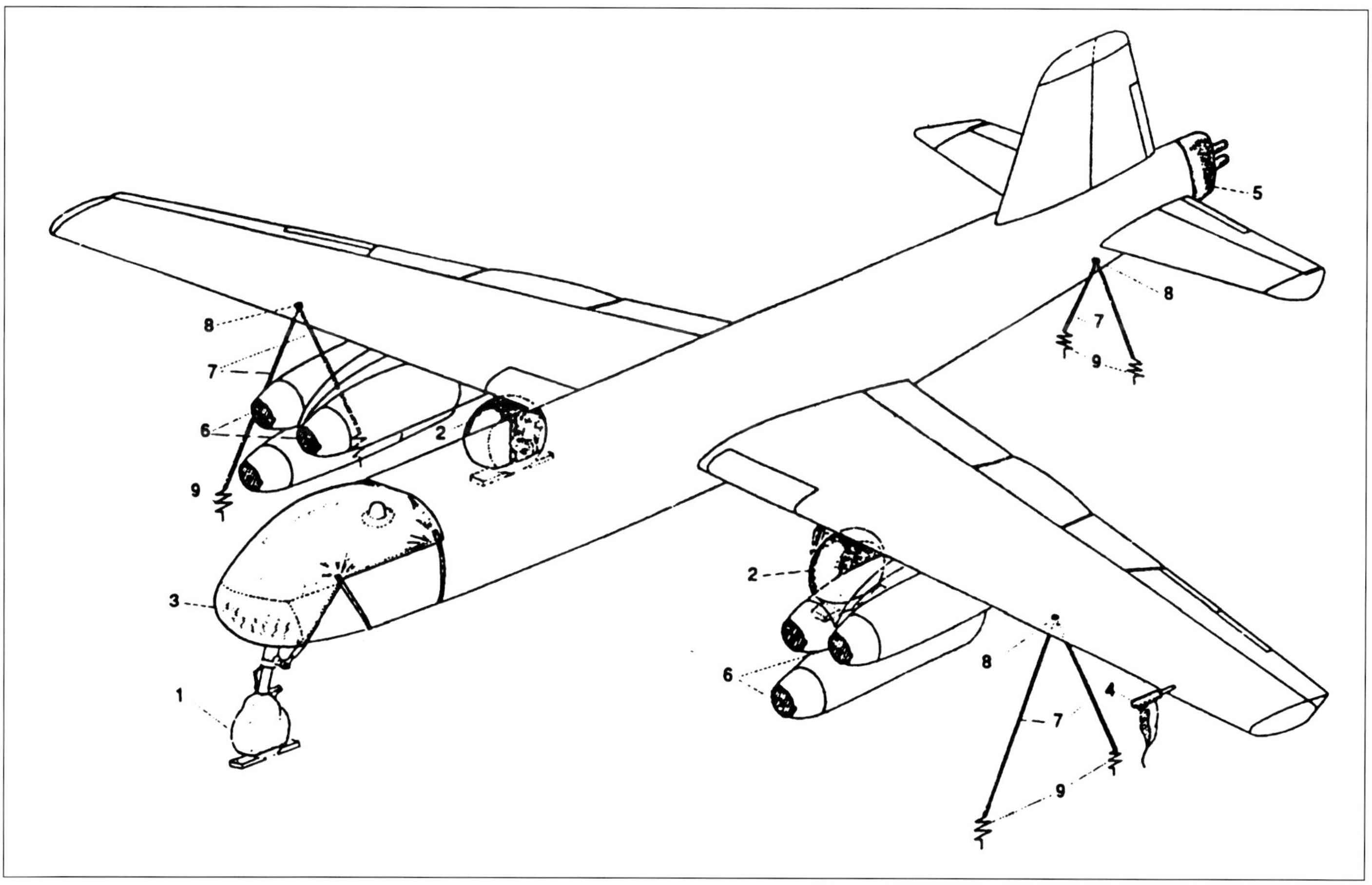

Sketch showing the covering and tie-down methods for the EF 131 (excerpt from the EF 131 service manual by Dr. Peter Korell).
Key:
1. nosewheel cover
2. mainwheel covers
3. cockpit cover
4. pitot tube cover
5. tail gun cover
6. engine covers
7. tie-down chains/cables
8. tie-down eyelets
9. ground anchor stakes

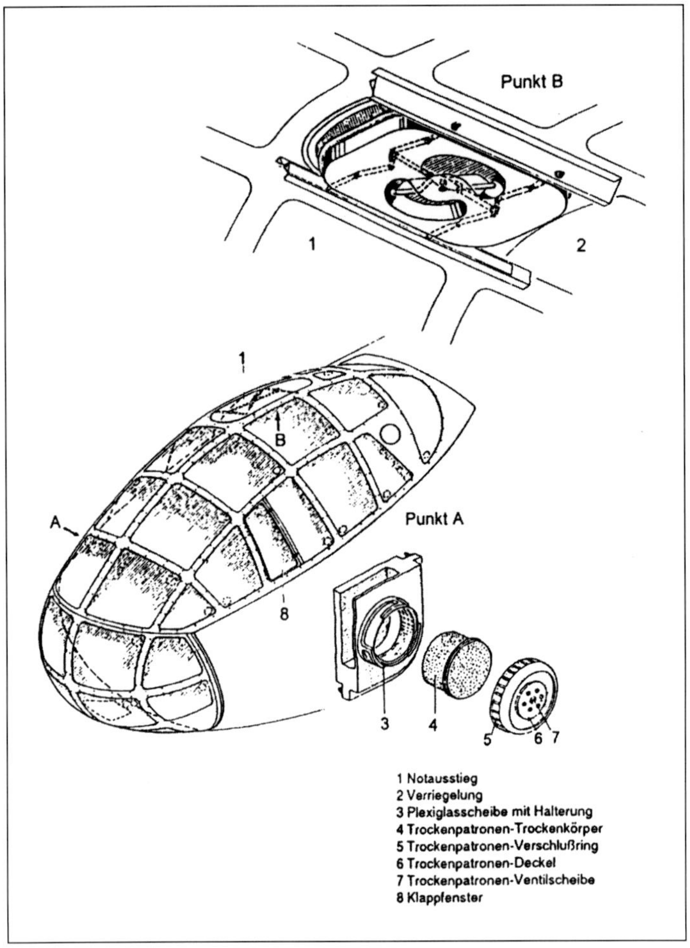

perature had dropped to as low as -35%; they simply would not start. Using a few tricks, such as giving them a powerful blast of ether, the Jumo 109-004s could be coaxed to life in most cases. For several months, snow drifts prevented any further flight operations. The EF 131 stood buried under a mountain of snow for quite a while, dormant on the test field. When winter finally departed, it was discovered that frost and moisture had taken their toll on all rubber parts as well as parts of the electrical system, and that all these needed replacing. As was the Russian mentality, this took considerable time, so that it wasn't until the first part of June 1948 that the EF 131 was again cleared for flight testing. But it was not to be. On 21 June 1948, the order came down from the highest levels of the Ministry for Aviation Industry (Order No. 440) directing that all activities with regard to the EF 131 be stopped.

With the success of a new, more powerful generation of engines the triple combination powerplant had outlived its usefulness, and with these new engines other projects were developed based on the EF 131. These included the EF 140 and EF 140R, which were also built and successfully tested. Since these aircraft unequivocally owed their existence to the EF 131 and, by default, to the Ju 287, they are briefly mentioned here, although a more detailed examination is not forthcoming since these airplanes were not born on German soil and are therefore purely a postwar development. Thus ended, quite undramatically, what had begun so full

Above: Sketch showing the canopy parts and emergency escape hatch with seals and hygroscopic element (excerpt from the EF 131 service manual by Dr. Peter Korell).

Key:
1. emergency exit
2. seal
3. plexiglass plate with mounting support
4. hygroscopic cartridge container
5. hygroscopic cartridge locking ring
6. hygroscopic cartridge cover
7. hygroscopic cartridge vent plate
8. hatch window

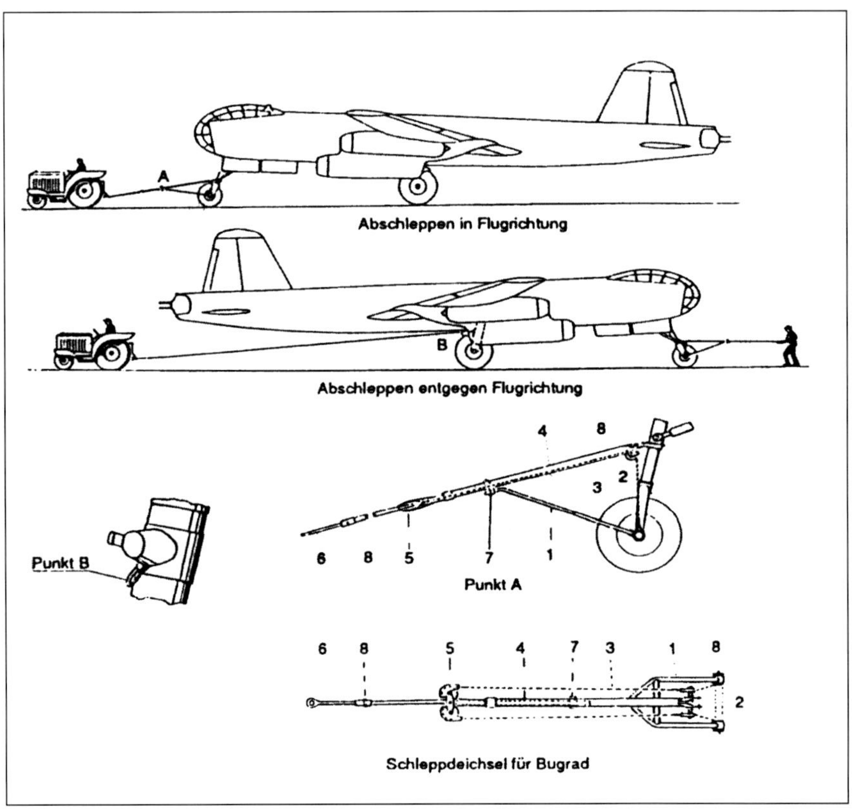

Right: Sketch from the EF 131 manual - towing using a tractor.
Key:
1. guide shaft
2. cable support pulley
3. tow cable
4. spindle
5. crossbar with pulleys
6. tow cable
7. adjusting shackle
8. slip bolt

Above left: The only known photograph of the EF 131 V1 shows the aircraft in the spring of 1948, parked in the open amid the melting snow. Above right: An enlargement shows that there was no armament in the tail barbette and that the fuselage terminates abruptly. Right: This detail magnification does not reveal any periscope sight in the upper cockpit area; most probably it was removed prior to the onset of winter.

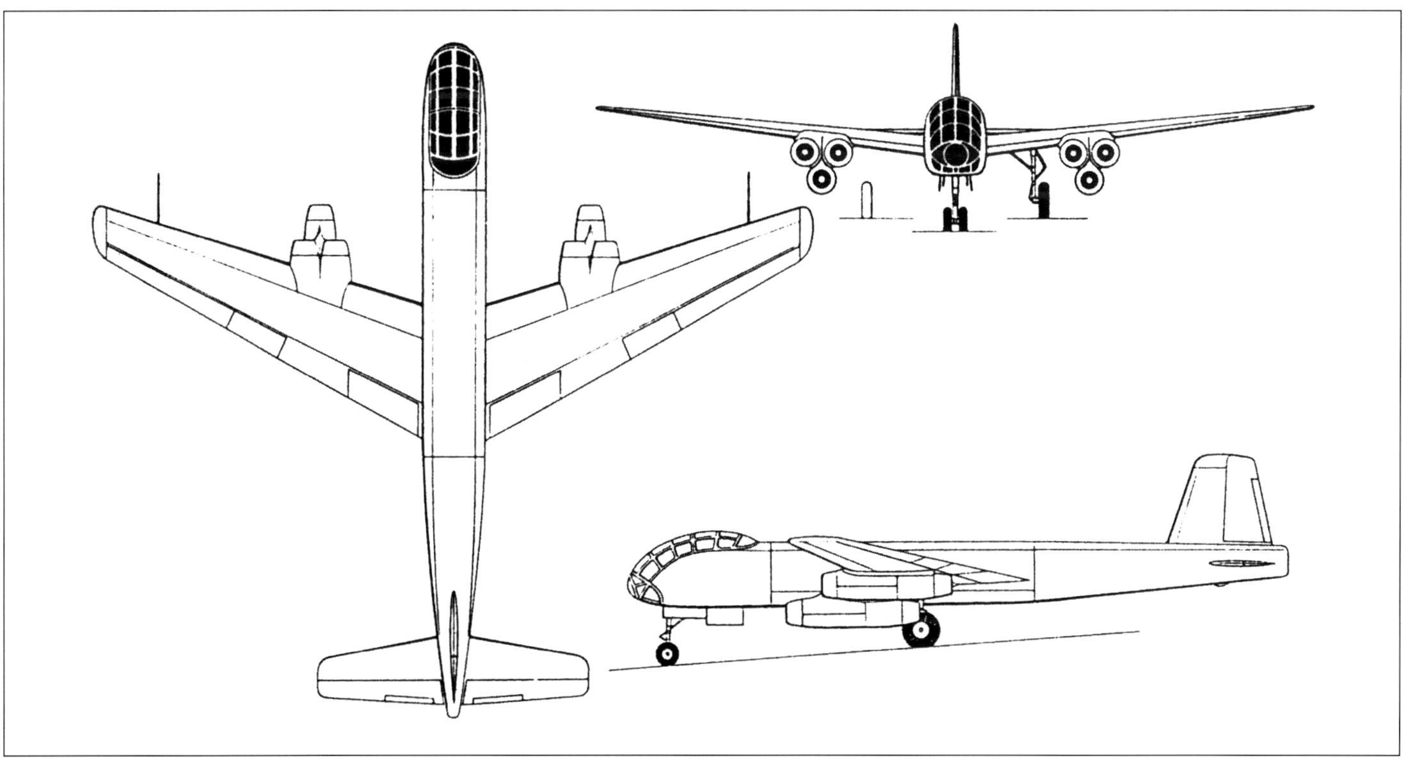

Three-view of the EF 131 V1 which, given the above photograph, is most likely how it must have looked.

Cover for the EF-131 guide by Dr. Peter Korell. The following excerpts are all taken from this handbook and speak for themselves through their drawings and text.

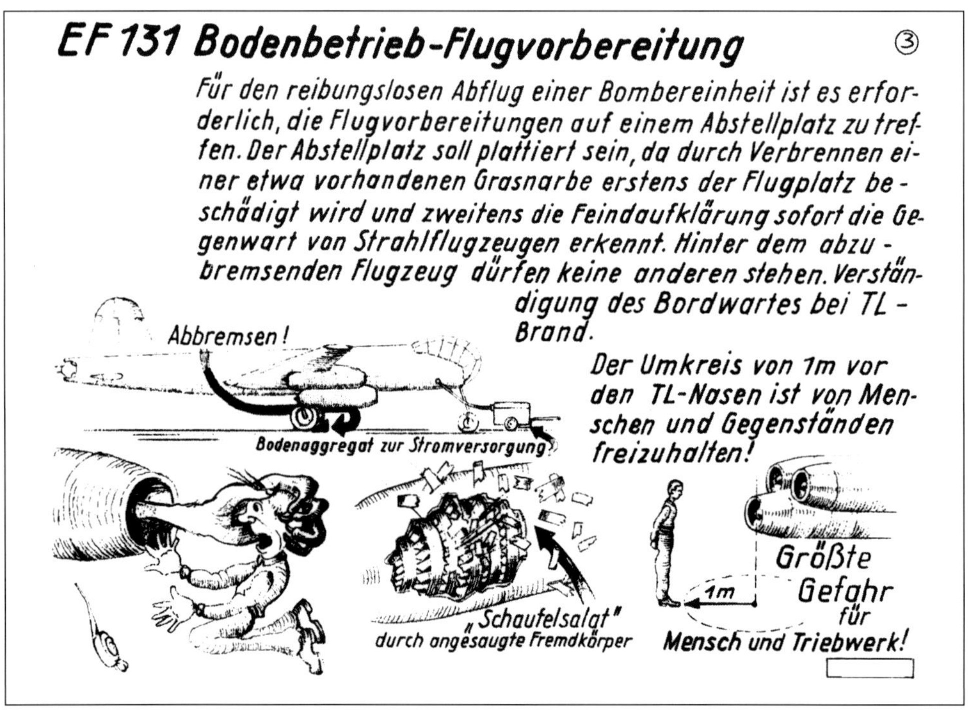
EF 131 Bodenbetrieb-Flugvorbereitung

3

Für den reibungslosen Abflug einer Bombereinheit ist es erforderlich, die Flugvorbereitungen auf einem Abstellplatz zu treffen. Der Abstellplatz soll plattiert sein, da durch Verbrennen einer etwa vorhandenen Grasnarbe erstens der Flugplatz beschädigt wird und zweitens die Feindaufklärung sofort die Gegenwart von Strahlflugzeugen erkennt. Hinter dem abzubremsenden Flugzeug dürfen keine anderen stehen. Verständigung des Bordwartes bei TL-Brand.

Der Umkreis von 1m vor den TL-Nasen ist von Menschen und Gegenständen freizuhalten!

Advice for ground servicing.

of hope in 1944/45 Germany. An appropriate résumé can be found at the end of this book, in the chapter entitled Epilogue and Credits, after page 214. It should be noted that, with regard to the EF 131, a 56 page thick "EF 131 Primer" was published in 1946 under the direction of Hermann Schmidt-Stiebitz, though it was never used in practice - something which does not detract from its interest, however. Dr. Peter Korrell has reworked the EF 131 Primer into a printed facsimile, with excerpts shown here. Those who have found this to their taste can purchase the entire 56-page thick EF 131 Primer directly from him (Dr. Peter Korrell, Cranachstraße 33, 38300 Wolfenbüttel).

Without sounding supercilious, the technical development in the area of supersonic projects is generally recognized as another bright gem in the crown of German engineering art, and will be illuminated further in our look at the Ju 287's competitors.

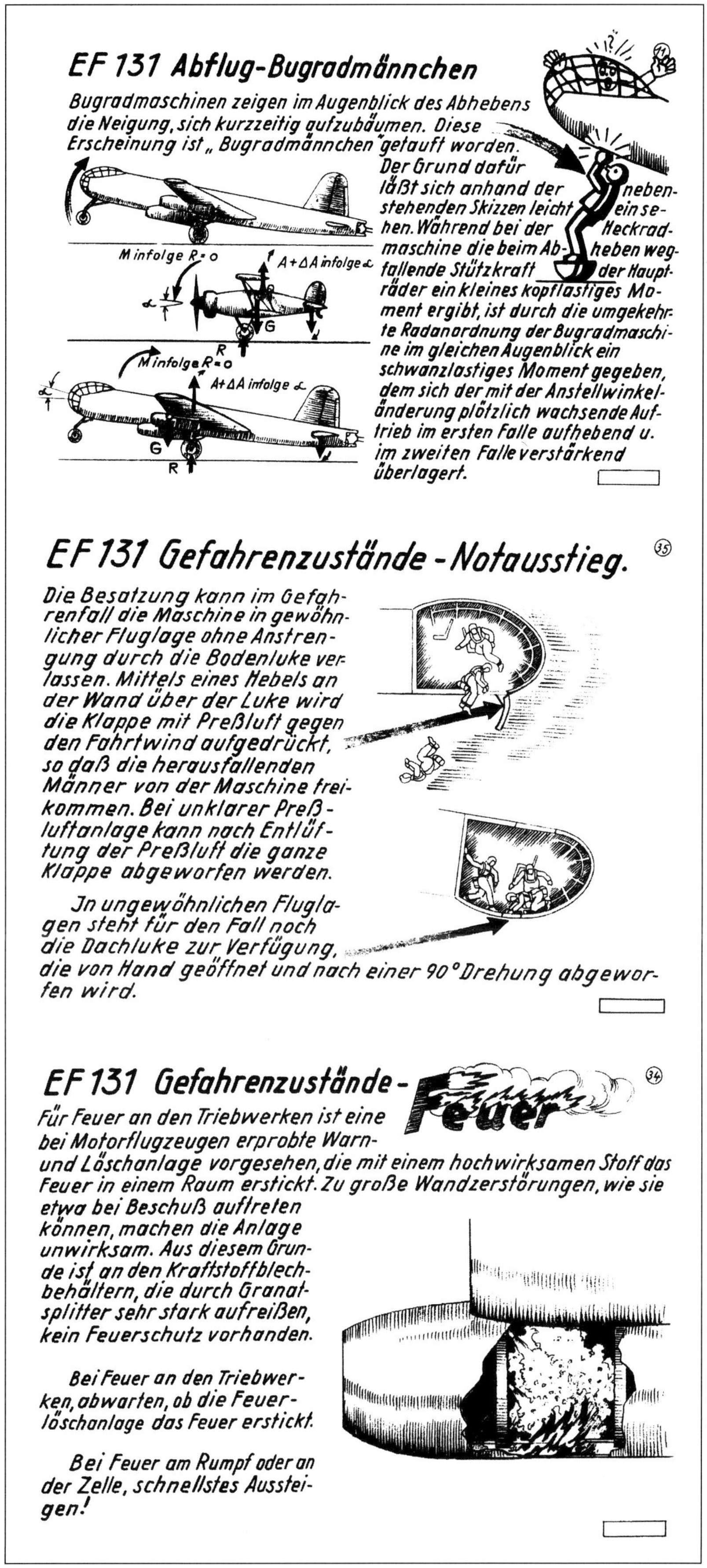

EF 131 Abflug-Bugradmännchen

11

Bugradmaschinen zeigen im Augenblick des Abhebens die Neigung, sich kurzzeitig aufzubäumen. Diese Erscheinung ist „Bugradmännchen" getauft worden. Der Grund dafür läßt sich anhand der nebenstehenden Skizzen leicht einsehen. Während bei der Heckradmaschine die beim Abheben wegfallende Stützkraft der Haupträder ein kleines kopflastiges Moment ergibt, ist durch die umgekehrte Radanordnung der Bugradmaschine im gleichen Augenblick ein schwanzlastiges Moment gegeben, dem sich der mit der Anstellwinkeländerung plötzlich wachsende Auftrieb im ersten Falle aufhebend u. im zweiten Falle verstärkend überlagert.

EF 131 Gefahrenzustände - Notausstieg.

35

Die Besatzung kann im Gefahrenfall die Maschine in gewöhnlicher Fluglage ohne Anstrengung durch die Bodenluke verlassen. Mittels eines Hebels an der Wand über der Luke wird die Klappe mit Preßluft gegen den Fahrtwind aufgedrückt, so daß die herausfallenden Männer von der Maschine freikommen. Bei unklarer Preßluftanlage kann nach Entlüftung der Preßluft die ganze Klappe abgeworfen werden.

Jn ungewöhnlichen Fluglagen steht für den Fall noch die Dachluke zur Verfügung, die von Hand geöffnet und nach einer 90° Drehung abgeworfen wird.

EF 131 Gefahrenzustände - Feuer

34

Für Feuer an den Triebwerken ist eine bei Motorflugzeugen erprobte Warn- und Löschanlage vorgesehen, die mit einem hochwirksamen Stoff das Feuer in einem Raum erstickt. Zu große Wandzerstörungen, wie sie etwa bei Beschuß auftreten können, machen die Anlage unwirksam. Aus diesem Grunde ist an den Kraftstoffblechbehältern, die durch Granatsplitter sehr stark aufreißen, kein Feuerschutz vorhanden.

Bei Feuer an den Triebwerken, abwarten, ob die Feuerlöschanlage das Feuer erstickt.

Bei Feuer am Rumpf oder an der Zelle, schnellstes Aussteigen!

Advice for action in the event of engine fire and operation of the fire extinguishing system.

EF 131 Bodenbetrieb-Aufstellung zum Abflug. ⑥

Aufstellung der Einheit bei schmaler Startbahn.

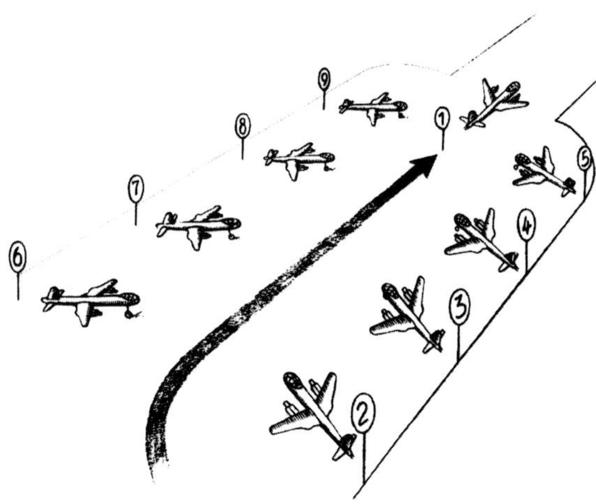

Gleichzeitiges Anlassen aller Maschinen.
Nach dem Anrollen des Staffelführers rollt der nächste in die Abflugrichtung.
Nach Gradrichtung des Bugrades bremsen und Vollast der TL.
Dann erst Bremsen lösen.

EF 131 Einsatz-Abwehrsicht. ㊸

Die Forderung schnell und hoch zu fliegen, bringt mit der Vereinigung der Besatzung in einem aerodynamisch glatten Druckraum für den Richtschützen des fernbedienten Heckstandes Erschwernisse mit sich. Während zum Absuchen des oberen Raumes hinter der Maschine nach feindlichen Jägern das bloße Auge genügt, muß der Schütze für den von der Maschine verdeckten unteren Raum das Periskop zu Hilfe nehmen. Das Periskop folgt über eine Rückmeldewelle den Bewegungen des Waffenstandes. Wegen einer besonderen Schnellflugverkleidung des Waffenstandes ist es erwünscht, möglichst lange auf dem Fluge die Waffe in Nullstellung zu lassen, wobei mit dem Periskop ein Kegel von 40° beobachtet werden kann. Beim Fahren des Waffenstandes und Absuchen des übrigen unteren Raumes fällt die Waffenverkleidung ab und der Maschinenwiderstand erhöht sich entsprechend den Stellungen.

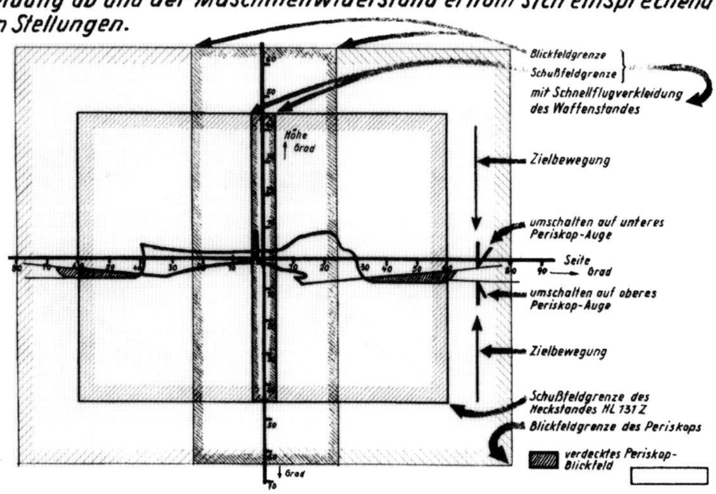

Above: Guidance for multi-plane operations; takeoff position and order. Left: Advice for fending off enemy aircraft in combat.

Junkers Ju 287 Competitors

The following chapters will illustrate the officially recognized developments that were in competition with the Junkers Ju 287. The Heinkel He 343 has repeatedly been classified as such. It is now high time that this cliché is dispensed with once and for all. The He 343 was never a competitor of the Ju 287. The He 343 covered a completely different tasking spectrum than the full-blooded Ju 287 turbojet bomber. The He 343 was only a competitor insofar as it was part of a decision on whether Germany could still afford the development of a pure turbojet powered bomber platform or would have to branch out into a multi-role turbojet aircraft such as the He 343. Anything else is not up for debate!

Actually, this would be one reason not to discuss the He 343 any further, particularly since there was the fact that the He 343 was not originally intended to be designed with a swept wing - another reason why it requires no further mention. But since there's been so much talk about the He 343, a separate chapter has been dedicated to this type.

Given the precarious situation facing Germany in 1944/45, it is no wonder that the Ju 287's competition was rather thin on the ground. The project studies of Messerschmitt remained just that - project studies. The flying wing turbojet bomber project by the Brothers Horten required massive support of the SS leadership that was increasingly finding its way to the forefront of aeronautics. This, however, does nothing to change the fact that the development and the first timid steps toward construction of this high-tech product of German engineering art deserve serious attention.

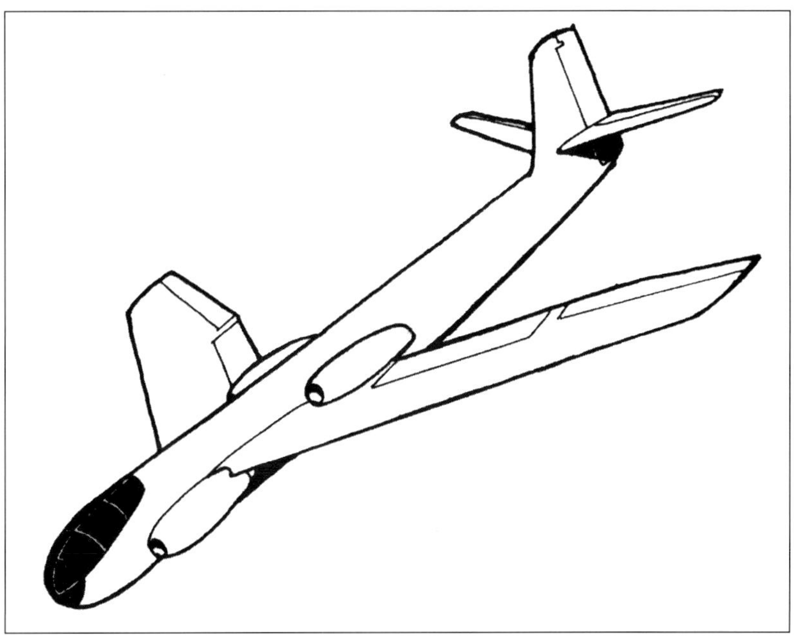

Project sketch of a late version Heinkel He 343 with swept wings and jet engines located on the forward fuselage as originally envisaged for the Ju 287, as well as wing-mounted engines.

Messerschmitt Me P1107 and Me P1108

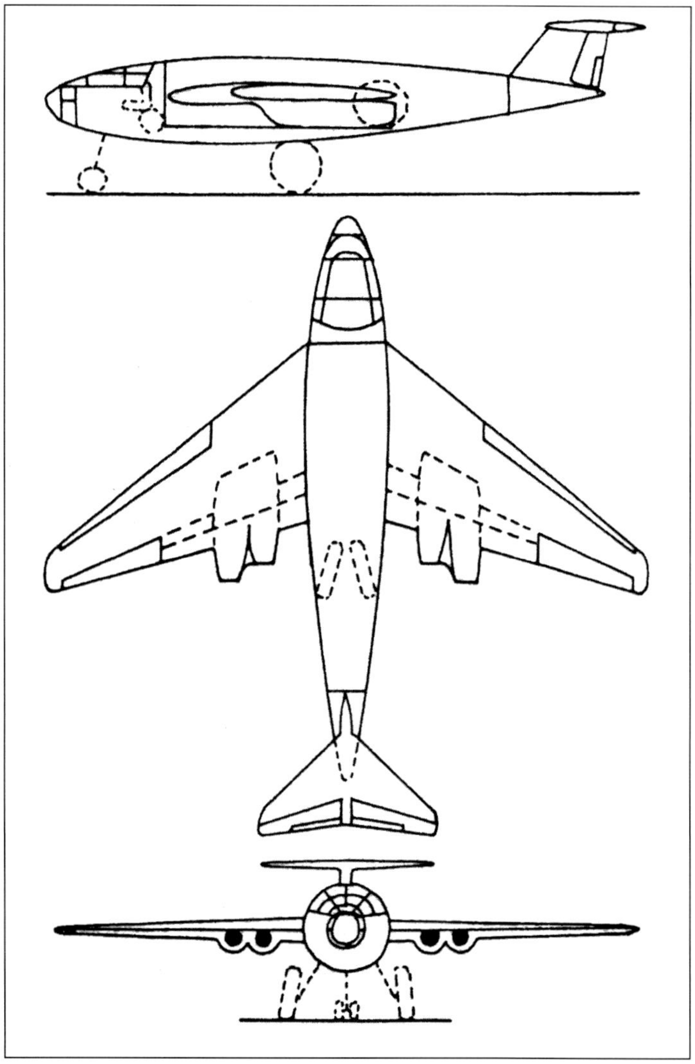

Early project drawing of the Messerschmitt Me P 1107-1 with T-tail and a relatively large fuselage diameter.

In the spring of 1945 the Messerschmitt company became involved in several studies for competing types to the Junkers Ju 287 S, the long range version of the Ju 287 (see page 50). Several design sketches were prepared under the project heading of Me P 1107 and Me P 1108, with these reflecting both normal fuselage construction as well as a pure flying wing design or delta configuration. All were the same in that they had swept wings, were powered by four Heinkel He S 011 jet engines and were designed for long range.

A few of these studies were of a very modern conception and therefore deserve a certain amount of attention; in this or that study one can almost see some of the postwar aircraft designs. A few publications have shown these sketches in a refined form or with more details and have passed them off as original company blueprints. The fact is, that these were never anything more than studies. The Messerschmitt company never was awarded a contract for constructing any of these projects. Professor Dr.-Ing. Günther Bock confirmed this in his testimony on the state of Germany's aircraft industry at the war's end before a commission of Soviet officers in September 1945.

In the spring of 1945 the Messerschmitt company carried out several studies for a long-range, four-engine bomber to compete with the Junkers company and the design from the Brothers Horten. It was clear from the outset that only a portion would and could be accepted for construction.

As pointed out in the next chapter, the Brothers Horten's design was heavily sponsored by the SS and it was inevitable that a corresponding contract for the construction of the type would follow in March of 1945. This, then, begs the question of how exactly the built-up version of the Me P 1107 or Me P 1108 might have looked had not the entire development been frozen in the project study stage. Nevertheless, what follows is a project study for both the Me P 1107 and the Me P 1108, representing the many other types drawn up.

Brief description of the Me P 1107 as ultra high-speed bomber with conventional fuselage configuration

Layout:	Oval fuselage cross-section as well as overall fuselage design, taken from the Me 264, empennage designed as V-type with slight positive sweepback. The wings, with a sweepback of 35%, are made of wood and carry part of the onboard fuel supply in protective wing tanks
Wingspan:	17.30 m
Length:	18.00 m
Height:	5.40 m
Crew:	Three men in heated pressurized cockpit for operations at higher altitudes
Engine system:	Four Heinkel He S 109-011 jet turbines, each with 1,300 kg static thrust
Armament:	4,000 kg payload, no further details
Maximum speed:	935 km/h at 7,000 m altitude
Cruising speed:	850 km/h at 8,000 m altitude
Range:	4,430 km
Service ceiling:	14,000 m

The description of the somewhat slimmer Me P 1108 shows the close relationship to the Me P 1107 and is unmistakable as a direct follow-on development.

Brief description of the Me P 1108 as ultra high-speed bomber with conventional fuselage configuration

Layout:	As Me P 1107, but with overall somewhat more slender and more stretched fuselage, empennage as per Me P 1107, designed as V-type with slight positive sweepback. The wings, with a sweepback of 35%, are made of wood and carry part of the onboard fuel supply in protective wing tanks
Wingspan:	19.80 m
Length:	18.20 m
Crew:	Four men in heated pressurized cockpit for operations at higher altitudes
Engine system:	Four Heinkel He S 109-011 jet turbines, each with 1,300 kg static thrust
Armament:	4,000 kg payload, no further details
Maximum speed:	980 km/h at 7,000 m altitude
Cruising speed:	890 km/h at 8,000 m altitude
Range:	7,000 km
Service ceiling:	14,000 m

It's not difficult to see how the two projects bore close similarity to each other, and it can be assumed that if a contract had been issued it would have undoubtedly been in favor of the Me P 1108, whose performance far outpaced that of any other relevant Messerschmitt project. But now let us leave the realm of pure speculation and delve once again back into reality, namely the quite interesting story of the Horten Ho XVIII.

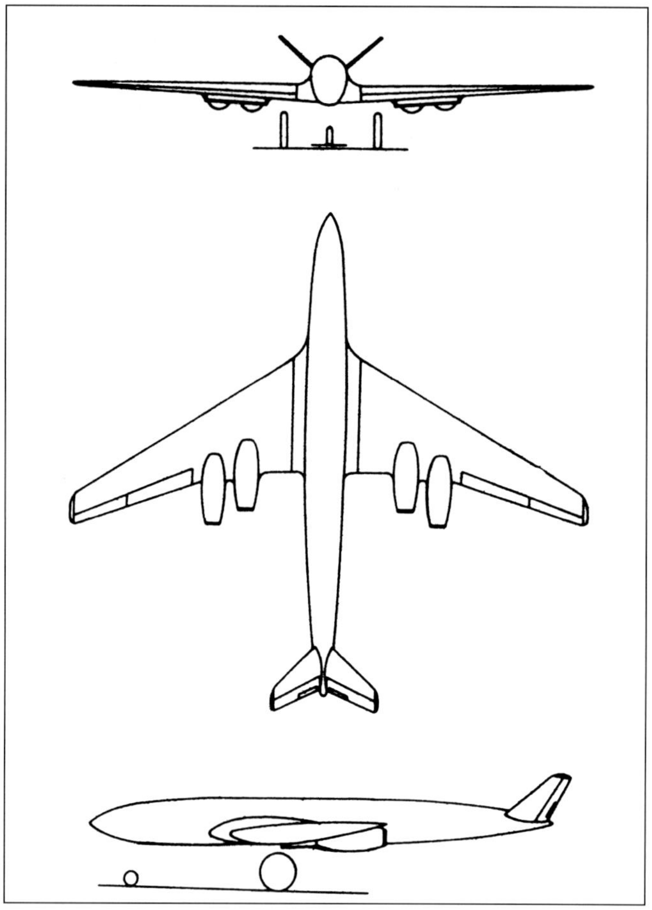

One of the first rudimentary project sketches for the Messerschmitt Me P 1108 with V-tail

Drawing of the Messerschmitt P 1108 by Jennifer Lommel as it would probably have looked had it actually been built.

Horten Ho XVIII

The competing development to the Ju 287 created by the brothers Walter and Reimar Horten can undoubtedly be classed as one of the most interesting designs ever. Even though very little is known about this design, the flying wing configuration and construction of the aircraft designated the Horten Ho XVIII were at extremely advanced stages at the war's end. Without a doubt, this was in large part due to the fact that this undertaking found strong support within the ranks of the SS, as we will soon discover. As mentioned earlier in passing, the SS took great interest in and had considerable influence upon the development of advanced technologies in the field of aviation. In a 7 June 1944 written memorandum - already cited briefly - from *SS- Standartenführer* Dr. Martin Brustmann to *SS-Standartenführer* Baumert, it continues as follows:

> ...On the other hand, the Americans have long known that for the last five years Germany has had aircraft whose speed and payload capacity would exceed any existing designs, assuming our aviation industry would have recognized the potential in time. The Americans are already building a heavy lift aircraft based on this principle of a flying wing, with a 120 ton takeoff weight and 40 ton payload (This undoubtedly refers to the development of the B-35 Flying Wing, a four-engine flying wing built by the Northrop company in the US, which flew for the first time in 1946 - author's note)! As a fighter, this flying wing aircraft's performance is superior to anything in existence when it comes to speed and climb rate; it is imperative that we push for the construction of such aircraft, and Dr. May, who has amassed experience in wing construction at his wood finishing plant in the Protectorate, is able to produce such aircraft in large quantities in a relatively short period of time. I myself have designed three aircraft types and would have had no success had not Prince Heinrich of Prussia, the brother of the Kaiser, sponsored me. Just as critical would be the protection of the SS with regard to the flying wing aircraft in order for us to obtain the potential and deserved success we so desperately need. I've been following aviation since 1911 and was able to recommend to Vice Admiral Dick, the organizer of naval aviation at the time, an aircraft that manifested this principle; now, too, I believe I have good reason to think that my recommendation of the flying wing concept will fall upon fertile ground. (...), the *Reichsführer* will be amenable to the idea of the flying wing design and create it using the circle of SS men and resources, something which will guarantee us the advantage in the air we so urgently need. (...) Please excuse this lengthy letter, but we cannot afford to lose any time in this matter....

That this letter had the desired effect is evidenced by a letter from *SS-Standartenführer* Paul Baumert, dated 22 June 1944, to *SS-Standartenführer* Dr. Rudolf Brandt:

Dear Rudi,

Attached I'm sending you an exchange of letters between the Minister of Aviation and the C-in-C of the Luftwaffe, where you can see that he wants to task *SS-Hauptsturmführer* Dr. May and *Ingenieur* Schaller with building twelve flying wing airplanes. I'm asking for the approval of the *Reichsführer-SS*. Since the matter is extremely urgent, I ask that this be submitted as soon as possible....

The prompt response came just a day later:

Dear Paul,

I'm returning the exchange to you. I wanted to speak with you on the telephone, but haven't had a chance. Construction of the aircraft is already underway, as I learned from a conversation with *SS- Obergruppenführer* Jüttner during a stopover in Bergwald. The *Reichsführer-SS* is in agreement....

Thus, the way was finally paved for construction of flying wings. Initially, the contract was for the design and construction of a little brother to the future Ho XVIII: the Ho IX, a twin engined flying wing fighter and fighter bomber (later the Go 229). So that pilots of future turbine powered flying wings could be trained on the unique handling of flying wing control, a contract was also issued for additional Horten Ho III flying wings. In a conference memorandum sent to SS Command Office, Office X (participants were *SS-Standartenführer* Hoffmann, *SS- Hauptsturmführer* Dr. May and *Ingenieur* Schaller of the Luftwaffe CinC test detachment), dated 15 June 1944, it reads:

In association with the many conferences that have been held on the subject named above, as of today's date a contract has been issued to the company of A. May, Stuttgart for twelve flying wing aircraft of the type Horten 3 (III) for training purposes and for ten flying wing aircraft of the type Horten 9 (IX) with turbojet engines for experimental purposes, later planned for use in large numbers in the fighter role.

In accordance with the agreement, six of the twelve aircraft (of the type Horten 3) are to be provided to the *Reichsführer-SS* as courier aircraft.

The remark about the six Horten Ho IIIs serving as courier aircraft seems somewhat odd, since the Ho III was only a flying wing glider. Most likely this was a reference to the twin-engine Ho VII, which was erroneously cited as the Ho III. Be that as it may, it was the Horten Ho IX - with two prototypes built for testing - that

Still minus its engines, the Horten Ho IX (Go 229) is seen in front of the hangar in Göttingen.

offered initial flying experience with the turbojet powered flying wing concept. However, only the second prototype, the Horten Ho IX V2, was powered by two Jumo 109-004 B-2 jet engines; the Ho IX V1 flew exclusively as an unpowered glider. The thoroughly positive flight tests were carried out by pilots Heinz Scheidhauer and *Leutnant* Erwin Ziller. *Leutnant* Ziller was killed on 18 February 1945 during a test flight in the Horten Ho IX V2. The cause was most likely engine failure. H. P. Dabrowski, in Vol. 17 of the *Waffenarsenal* series, writes:

> Dropping from high altitude after about 45 minutes flight time, at about 800 m Ziller repeatedly pushed the airplane down and then pulled up again, apparently trying to restart a flamed out engine. Unexpectedly prematurely, the landing gear extended at an altitude of 400 m, the airplane became slower, banked with a howling engine sound and suddenly went inverted into a right hand spin banked at about 20%, faster and faster in three ever tightening circles. At the start of the fourth circle the machine struck the hard frosted grass outside of the airfield. The first man from the Horten group was Walter Rösler, who reached the scene of the crash after about 2 1/2 minutes. Here is an excerpt from his report:
>
> "The first thing I saw near the hedges on the other side of the road were the two Junkers engines, and I heard the turbine blades spinning in the still warm left engine, whereas nothing moved in the other cooled engine alongside it..."

The stench of fuel was everywhere, but there had been no explosion on impact. Except for the engines and the plexiglass canopy, the aircraft was totally destroyed. Like the engines, Ziller had been thrown from the machine, striking a thick fruit tree - he was killed immediately. He had not used the radio, did not try setting up a final approach on a single engine with gear retracted, did not attempt to use the ejection seat and parachute, nor was the cockpit canopy jettisoned; obviously he wanted to save the valuable plane. What had happened? The empty compressed air tanks found in the wreckage confirmed that, when the engine died and the hydraulics failed as a result, the emergency system for the undercarriage kicked in and extended the gear using compressed air. Did this cause the boundary layer to separate, starting at the outer starboard wing area? Was the test pilot possibly rendered unconscious, and thus unable to react, by the burning oil from the heated engine (which was not sealed off from the cockpit by any bulkhead)?

The Gotha Go 229 V3, also powered by two Jumo 109-004 B-2 jet engines, did not go into flight testing before the end of the war. The Americans captured the Go 229 V3 in nearly finished condition at the Gotha company's production facilities in Friedrichsroda. Today this highly interesting plane wastes away in the warehouse of the Smithsonian National Air and Space Museum at Silver Hill (Washington, D.C.) in the United States, but there's thought of restoring the aircraft in Germany in the near future. One can only hope that the efforts and negotiations to this end are positive.

Building on the experience of the Ho IX, the Ho XVIII was designed virtually as an enlargement of the Go 229 with six Jumo 109-004 B-2 engines. With these engines, it was expected that the Horten Ho XVIII would be propelled along at speeds of about 820 km/h, and its range of 12,000 km meant that it could easily reach New York non-stop - with a bomb payload in overloaded state of 4,000 kg (like the Ju 287). This flying wing would have had a positive sweepback of 24.3%, which was within one degree of the sweep on the Gotha Go 229. The three-man crew sat in a heated, pressurized cockpit, allowing for high altitude flights in order to keep the fuel consumption to a minimum. The Ho XVIII was to have been built almost completely of wood, with the added benefit that this would give it a minimal radar echo and making it virtually invisible to the radar equipment at the time. A true Stealth bomber, then.

On 25 February 1945 a complete and refined set of blueprints was submitted for the Ho XVIII and as early as 12 March 1945 the SS Command Office, Office X issued a contract for construction of the Ho XVIII (other sources mention that Göring was the one who issued the contract on 23 March 1945. This is highly improbable, since by this time Göring had long since been stripped of his powers and would therefore have been unable to make such far-reaching decisions.). Against the background of these unverifiable reports there appeared in the war journal of the "Chief, TLR", page 147, for the reporting period 16 March to 4 April 1945, the laconic remark:

> **Horten Bomber Project:**
> Opinions and thoughts can be found in the included memorandum.

This leads one to assume that in fact no contract had been issued by this time. On the contrary, an actual contract wasn't issued until 12 April 1945 at the behest of the SS. Furthermore, production startup date was given as 1 April 1945 at Kahla near Weimar. Completely contradicting this, however, is a letter from 16 June 1944 to *SS-Standartenführer* Dr. Kurt May:

The Ho IX V2 was powered by two Junkers Jumo 109-004 B engines, seen here preparing for a jet powered test flight in Oranienburg. One of the ground crew must lie on his belly in order to help test pilot *Leutnant* Erwin Ziller fasten his safety harness.

The completed but never flown Go 229 V3, minus its wings, seen shortly after being captured by the Americans. The oversized nose wheel stands out quite well in this photo.

Another photograph of the Go 229 V3 from the port side reveals to good effect the Jumo 109-004 B engines buried in the wings. The engines were to have been buried in a similar fashion in the wings of the Ho XVIII.

Above: Reconstructed in-flight drawing of the Ho XVIII, with guns in the nose and aft section of the crew compartment. This corresponds generally to the two-view drawing on the following page. Below: Data table for the planned Ho XVIII with data for the originally planned six Junkers Jumo 109-004 B engines:

The company of A. May, furniture factory, Stuttgart, which was converted at the beginning of the war into a subcontractor manufacturing wings for the Messerschmitt Works in Augsburg, is recognized as an independent developmental works for the H. program (author's note - H. program refers to the Horten program). Its operation is confirmed as having "primary responsibility" with regard to this production program.

The intent is to assign you personally as the official in charge of startup and production of the H. program. It is requested that the approval of the *Reichsführer-SS* to this appointment be obtained as soon as possible. ...

Since this appointment followed immediately, *SS-Standartenführer* Dr. Kurt May, who had his office at the SS Command Office, Office X in Berlin-Wilmersdorf, Kaiserallee 188, from that point onward became responsible for the production interests of the Gotha Go 229 and, from the spring of 1945, for the Horten Ho XVIII as well. As was common for the SS whenever it had its finger in a

		Horten H XVIII	
Role		Long-range bomber	
Construction	Fuselage (center section)	steel tubing	
	Wings	wood	
Crew		three persons	
Engine number, type, version		six Jumo 004 B turbines	
	Rated thrust daN	6 x 900	
Wingspan		m	40
Leading edge sweep angle		degree	24.3
Taper ratio		8.0	
Root thickness		%	16
Root chord depth		m	6.5
Rib spacing		m	0.4
Wing area		m^2	150
Aspect ratio		10.7	
Pilot position			seated
Width, center section		m	4.0
Width, cockpit		m	1.6
Height, cockpit		m	1.2
Weight, empty		kg	11000
Ballast (weapons, ammunition)		kg	700
Load (payload)		kg	4000 (300 for crew)
Fuel		kg	16000
Max. flight weight		kg	32000
Wing loading		kg/m^2	213
Power loading		kg/daN	5.93
Stall speed		km/h	192
Landing speed		km/h	136 (16 tons)
Cruise speed		km/h	750
Max. speed (level flight)		km/h	820
Not to exceed speed		km/h	900

Air-to-air photo of the American YB-49, which was also powered by six jet engines and first flew in October 1947. As it was based on the same concept as the Ho XVIII, the picture shows to good effect what the Ho XVIII might have looked like if it had been built.

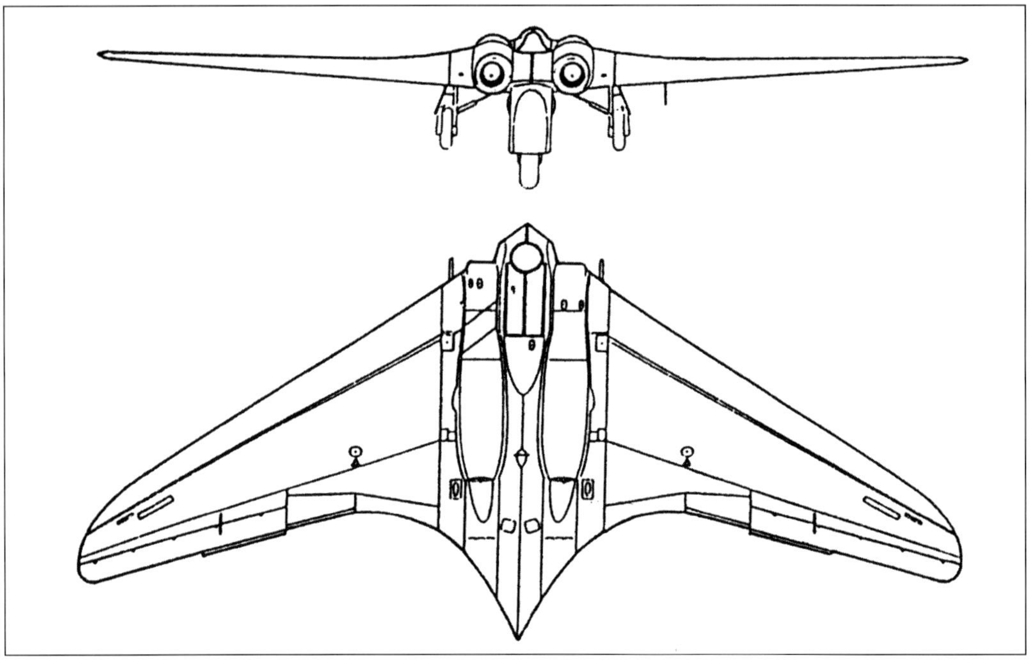

Two-view drawing of the Horten Ho IX (Ho 229, Go 229)

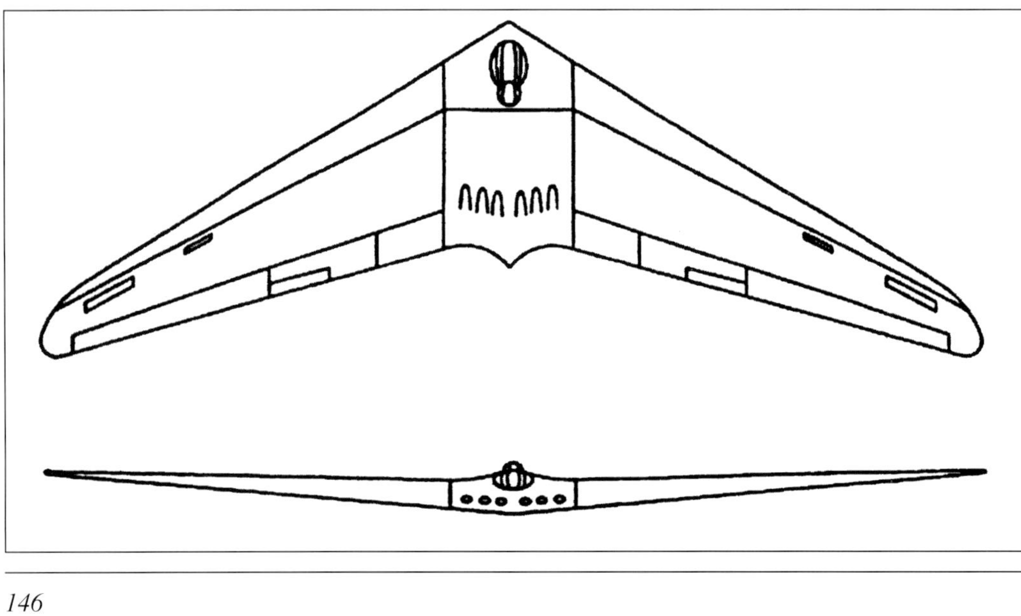

Two-view of the Horten Ho XVIII. A comparison of the two drawings reveals the close relationship to the Ho IX.

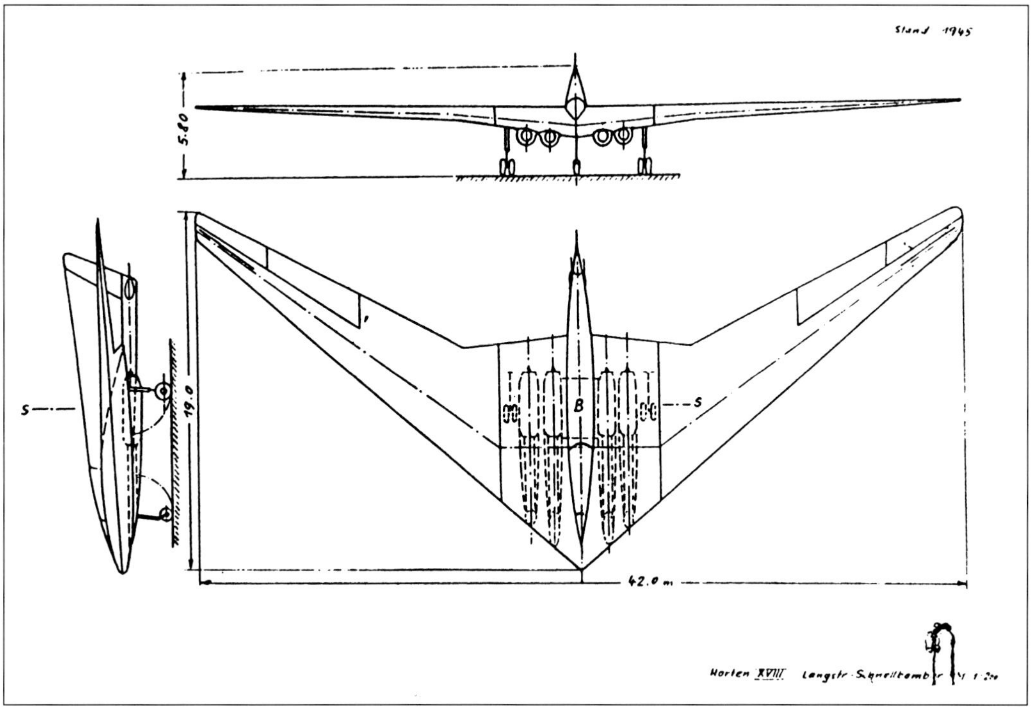

Three-view drawing of the Ho XVIII with a single central rudder. This drawing was reconstructed after the war's end and shows a four-engined variant with four He S 011s as powerplants.

particular pie, a figurehead was needed in the foreground for the H. program. *Oberst* Siegfried Knemeyer, who had already overseen the Junkers Ju 287 STRABO project as member of the technical armaments staff, was selected for the Ho XVIII bomber as "project officer".

Work on setting up the production facilities at Kahla proceeded at a snail's pace, and to compound matters they were mainly used for production of the Me 262. This meant that it was unlikely that production ever started on the Ho XVIIII at that site. What is more probable is that this - as planned - took place at the May Company's production facilities in Stuttgart. Be that as it may, there is insufficient documentation for either type to confirm this hypothesis. The only fact is that production did indeed begin before the end of the war in whatever capacity that may have been. Thus ended this promising development as well, just like Beethoven's "Unfinished" symphony - never completed. The claims that one example of the Ho XVIII was indeed finished before the war's end belong to the land of fairy tales and legends.

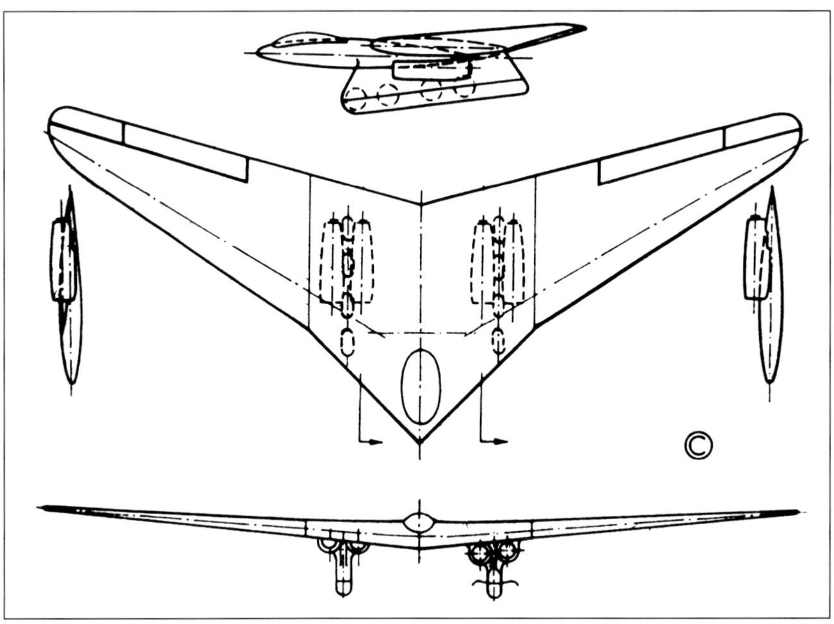

Just to add to the confusion: another three-view sketch of the Ho XVIII with four He S 011 jet engines and massive shrouds for the landing gear. The latter serves as additional stabilization and something of a replacement for the vertical stabilizer. Note the asymmetric configuration of the engines. Which of the depicted versions would have entered production is pure speculation.

Heinkel He 343 (He P1068)

The history of the Heinkel He 343 is as checkered as it is contradictory and shrouded in mystery. Only a handful of original documents exist and just a few of these have filtered out from the Russian archives.

Russian forces pushed into Austria from the east in April of 1945, and by the 15th of that month had closed a ring besieging the city of Vienna. A few days later the Russians had occupied the Heinkel Works outside of Vienna. In early May 1945 two Russian engineers from the People's Commissariat for Aviation Industry, Bondaryuk and Savitayev, paid a visit to the Heinkel Works at Vienna-Schwechat in order to get an idea of the technical state of development for Heinkel's projects. Among the many interesting developments at the Heinkel Company, they found the completed wing sections as well as parts of the fuselage for the first prototype, the Heinkel He 343 V1; this has been confirmed by Russian sources that have only been made available within the last few months. This dispels the notion that "the He 343 was never built". Heinkel had apparently exercised his notorious pigheadedness and, despite a restriction on building the He 343, went ahead with its construction. Heinkel himself claimed at the end of the war that construction had been underway on as many as three prototypes, of which at least the He 343 V1 had been completed (something which cannot be verified based on current evidence). This takes us back to the year 1944.

In mid-August 1944 there was a meeting held between Messrs. Prof. Dr. Ernst Heinkel, Director Hayn, Director Franke, the designers Schwärzler and Butter as well as Mr. Müller from Sales. During the course of the meeting it was discovered that the original contract for 20 prototype variants, not to include an initial two that would be used for testing and a third for static experiments, still was valid as originally assigned. Furthermore, the material for these 20 prototype versions must be required at any cost. This meeting was recorded in an internal memorandum, TD 45, dated 16 August 1944. The folks at Heinkel firmly believed that the Ju 287 would 'be a flop' and would, as a result, have a stronger argument and marketing opportunity.

Radiating self-confidence, a quite accurate "Assessment of the Overall Situation" was compiled, which is reproduced word for word here below before going into the actual design features of the He 343 itself. The following report shows that Heinkel, instead of focusing on a description of its own product, believed that it was more important to attack the Ju 287, and as a result the reader will note Heinkel sacrificing any discussion of its own merits while highlighting the defects in the other type:

The contract issued for the Ju 287's production startup, beginning in early 1945, reveals an attempt to acquire desired types with large payload capacities both as a final remedy and as a quick solution.

C-E 2/II considers this attempt risky and stands by their recommendation to begin immediate production of the He 343 exclusively. 2/II has suggested that Ju 287 production startup, i.e. August/ September 1944, won't be without its problems, at least for the first aircraft.

The timetable drawn up by the JFM (Junkers-Flugmotoren-Werke) for the Ju 287 is considered by 2/II to be overly optimistic, given the fact that the blueprints for the type weren't completed until mid-'44.

A technological comparison shows that the He 343 in its current form has about the same payload and range as does the Ju 287: 2 metric tons at 1,600 km vice 2 metric tons at 1,700 km. And performance with regard to takeoff, climb rate, operating altitude and speed are also practically identical.

The payload options for bombs do not differ radically from each other. The He 343 has four engines compared to the six of the Ju 287. With regard to equipped weight, it is about 2.5 - 3 tons lighter than the Ju 287, and takeoff weight is 4 - 5 metric tons less. Fuel weight is 1.5 tons less. The He 343 covers the reconnaissance role cheaper than the Ju 287 and furthermore is also suited for the heavy fighter role.

The space potential of the Ju 287, which is currently seen as an especial advantage, actually is of no consequence at all in the long run - because of the Ju 287's flying weight restrictions necessitated by performance considerations.

The performance limitation of the Ju 287 is at about 23 tons and is derived from the fact that after takeoff the climb rate cannot drop below 2 meters per second (for safety reasons). This means that, after lifting off the runway, the type needs a distance of about 1,700 meters to clear an altitude of 20 meters. Any more shallow climb after takeoff would pose problems. Incorporated in these 23 tons are currently 7 tons of fuel and 3 tons of payload. If one factors in the ever increasing airframe weight (based on experience), plus with light armor for the fuel tanks (hitherto unprotected) as well as with additional equipment, then effectively the type has but 2 tons payload at maximum takeoff weight, full tanks and without armament. If armament is included as per the requirement by the G. d. K. (author's note: *General der Kampfflieger* - General of the Bomber Forces), this would then impact the payload weight even further. Since the current figures are based on an increased thrust of 900 kg, there can be no anticipation of further thrust performance improvement and consequent flying weight increase for the time being. This, then, results in the fact that, despite the spacious airframe of the Ju 287, this space cannot be utilized for payload anytime in the foreseeable future. The Ju 287's larger dimensions thus offer no appreciable advantage over the He 343 at the present, instead costing more in terms of engines and metal alloys.

The planned shift to He S 011 engines does not change this comparison in any significant way, since the engine thrust practically remains the same and only the engine weight is reduced. It is anticipated that a full-sized Ju 287 would require six He S 011 engines.

With regard to the airframe itself, a decision in favor of the He 343 would enable simplification with regard to the powerplants; there would be no need for the BMW turbines.

From the airframe's viewpoint, the engine situation is as follows:

1) The airframe wants to have the fewest types of engine as possible for simplification of installation. Primarily, this means only one engine type for each class in production, where possible.

2) The airframe desperately need the powerful He S 011 engines for the existing types and the BMW 018 for larger new developments (flying wing).

3) Little can be done with the 800 to 1,000 kg class engines; aircraft with one turbojet engine only offer reasonable performance with the He S 011.
With regard to the turbojet-powered airframes, the Arado 234 will soon need the He S 011 for improvement of general performance, and the Me 262 for reasons of improving takeoff and higher altitude performance. The bomber airframes also show improved performance with the He S 011, especially with regard to range.

4) The BMW turbine is a later development of the 800 to 1,000 kg class engine. It follows a year after the Jumo turbojet engine. Whereas the Jumo engine now functions flawlessly, the BMW turbine in the Arado 234 has not exhibited a single flight without faults as of now. It would be more cost effective to focus all efforts on the He S 011 with the goal of having this engine available instead of getting yet another engine in the same class as the already working Jumo turbine that, as a result of its limited development potential ultimately impacts on the urgently needed more powerful engine types. Dropping the BMW turbine would be sensible. For almost a year BMW has claimed that the engine was ready for flying; however, in spite of incessant efforts on the part of everyone concerned there have been no problem-free flights to date.
It is obvious that when it comes to engine planning there are many other important factors favoring the BMW turbine with regard to conservation of resources as well as engine development. Nevertheless, it is critical that this planning, when in conjunction with airframe planning, take into account a possible streamlining of the engine situation.

It is not hard to see that, in their hearts, the people at Heinkel hoped that Junkers would fall on its face with its project and would, as a result, enable Heinkel to advance to the fore. Professor Dr. Heinkel did not want to accept that any design from his company had so blatantly been relegated to the minor league. In addition, he grumbled about the fact that Professor Dr. Hertel, above all people was now a successful opponent of his. Hertel had left him standing with the unfortunate He 177 project. At the time he had been the technical director at Heinkel and played a significant role in the development of the Heinkel He 177, formerly the project He P 1041, when he had a falling out with Heinkel, gave his notice, and went to work for Junkers. One can see how the malicious dispute between Heinkel and Junkers, with a variety of personal attacks, played a not insignificant role. Another important point was the proper emphasis on the engine situation. Heinkel had quite correctly recognized that it was irresponsible, given the shortage of resources, to have several of the same developments running side by side in the field of powerplants, sucking up personnel and materials in a wasteful manner instead of utilizing these for future developments offering more potential. But let's return to technology.

Around the middle of 1943 the RLM called for a superfast, two seat bomber capable of carrying a bomb load weighing 2,000 kg over a distance of 1,000 km at a speed of over 800 km/h. The first project studies at Heinkel were carried out under the project number P 1068 and envisioned a four- to six-engine cantilever mid-wing design, which initially would be powered by four He S 011 Heinkel turbojet engines. With a wingspan of about 19 m and a length of almost 20 m, the machine was equipped with a relatively small tricycle landing gear having a track of 2.80 m, something which later led to criticism. Since Heinkel was already overburdened with other projects, some of the work was subcontracted out to the DFS (*Deutsche Forschungsanstalt für Segelflug, e. V.* - German Research Institute for Glider Flight) at Hörsching. This was a fairly common practice, and the DFS had already done preliminary work for the He 280, a twin-jet fighter. Initially, the DFS conceived of a wooden wing for the P 1068, but wasn't persuasive enough to convince others involved in the project. In the meantime, time was pressing, and someone came up with the idea of simply making the Heinkel design, which in its basic configuration resembled a larger Arado 234, virtually a 1.5x enlargement of the Arado 234. On this basis work continued on the Heinkel draft, which the RLM had now officially designated 8- 343. From this point onward, the project was also known within Heinkel as the "STABO 16", referring to the fact that the jet bomber layout was to be designed for 16 metric tons. It was soon discovered that the Heinkel He S 011 engine would not be available anytime in the immediate future, and since it appeared that the He 343 would indeed be built, Heinkel gritted its teeth and opted for the Junkers Jumo 004 B engine. In order to push into the transonic flight region, plans for later variants called for swept wings with a positive sweepback of 35%. With the typical "Heinkel tempo" work began on turning the project into reality. It even continued, albeit somewhat more slowly, secretly in spite of the official orders halting work on the He 343 and redlining it in favor of the emergency fighter program. Apparently, before the war's end assembly had begun on several prototypes as well as completion of design work on variants A through D, although none of these actually ever left the assembly lines at the Heinkel Works.

The model sketch of a Heinkel He 343 provides a good overview of how the aircraft probably would have appeared.

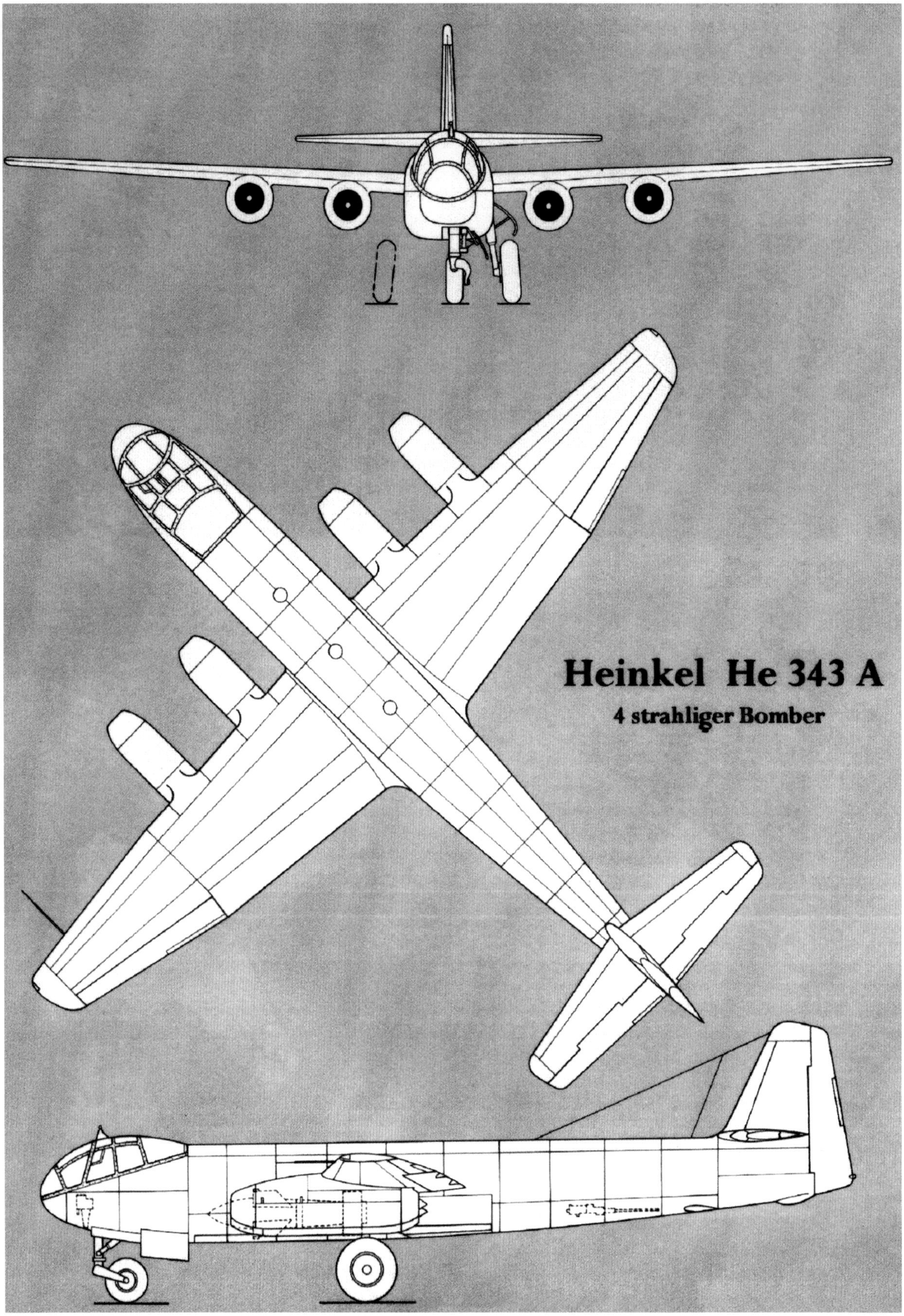

Three-view drawing of the bomber version of the Heinkel He 343.

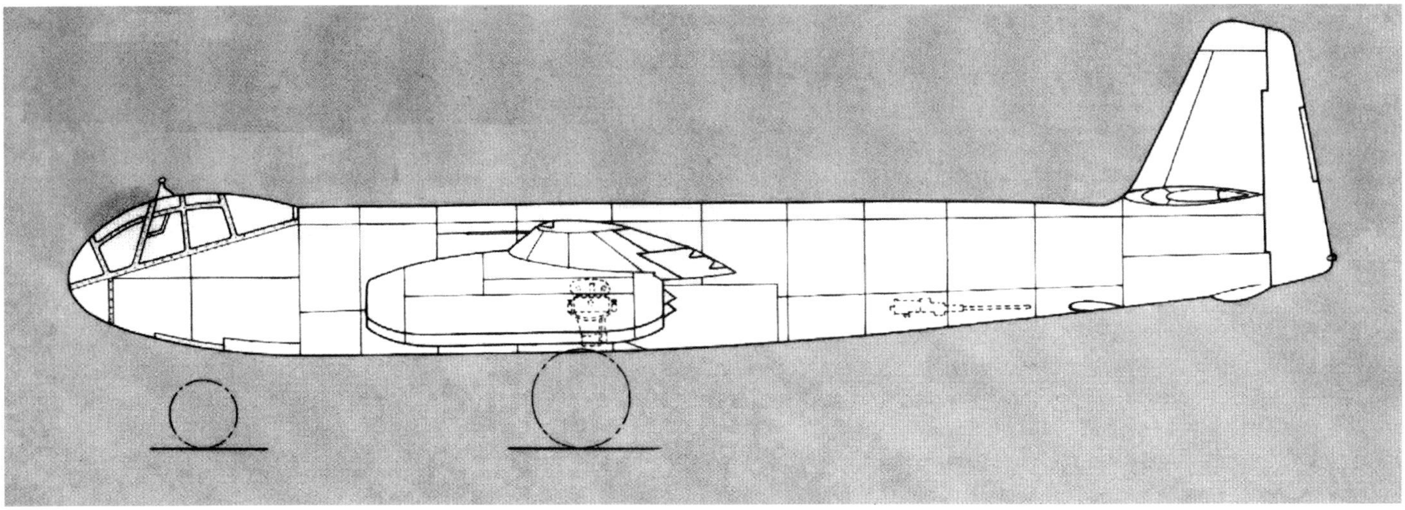

Reconnaissance version of the Heinkel He 343 with a scattering of cameras in the bomb bay.

He 343 A

Basic version planned in two variants as heavy fighter and high speed bomber

Operational version: four-engine bomber and heavy fighter

Wings: single-spar all-metal construction with Fowler flaps between the aileron and fuselage, with a wingspan of 18 m

Fuselage: monococque construction with oval cross-section, with a total length of 16.5 m, forward fuselage section fully glazed

Empennage: cantilever standard empennage of all-metal construction with trim tabs along the entire trailing edge of the control surfaces

Undercarriage: retractable tricycle landing gear, main tires retracting aft into the fuselage sides, nose gear also retracting aft and rotating through 90% into the fuselage nose

Engines: four Jumo 004 B (4 Jumo 004 C) on separate mounts beneath the wings, fuel carried in four fuselage cells for a total of 7,450 liters

Crew: two men in fully glazed nose compartment designed as pressurized cockpit

He 343 A1

Heavy fighter with armament consisting of four MK 108 machine cannons of 30 mm caliber, side by side in a gun bay beneath the fuselage nose, fixed, forward firing, plus two MG 151 20 mm machine guns at the base of the aft fuselage, fixed downward firing.

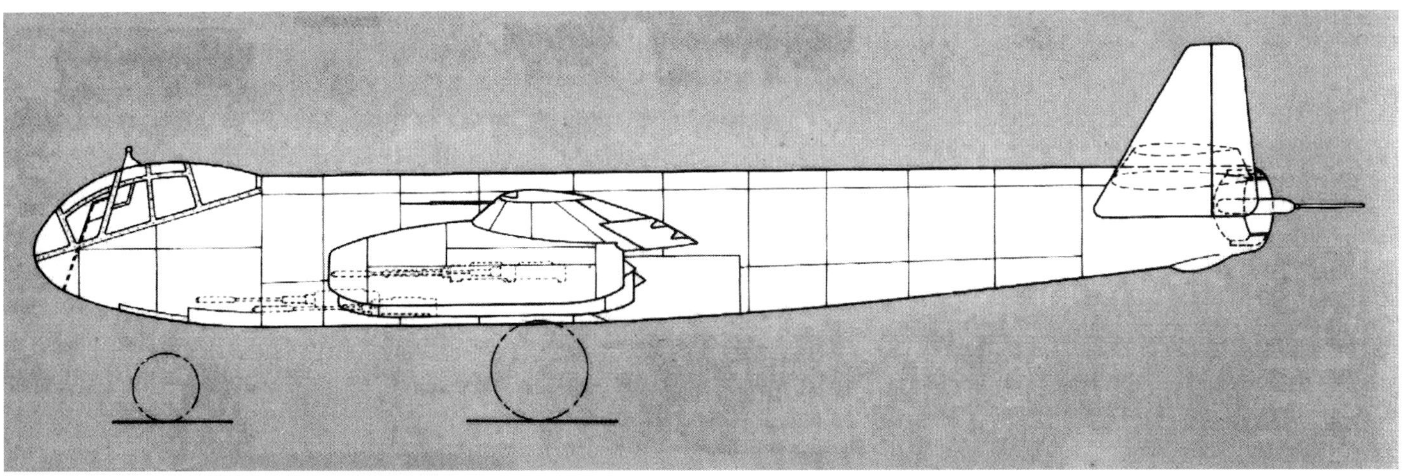

Heavy fighter version of the Heinkel He 343 with triangular twin rudders and remote-controlled tail gun position.

The postwar Russian design by and large was the same as the original Heinkel He 343 and offers a good impression of what the He 343 would have looked like.

He 343 A2

High speed bomber with free fall bomb payload totaling 2,000 kg carried internally in a bomb bay, plus two MG 151 20 mm machine guns at the base of the aft fuselage, fixed downward firing.

He 343 B

Slight improvement on the He 343 A1 as heavy fighter with same powerplant as well as nose armament, but a remote controlled turret in the tail, sighted via an FHL 151 Z periscope gunsight. The He 343 B series also was fitted with a two-section triangular rudder similar to that found on the Dornier Do 317.

He 343 C

Modification of the He 343 A2 with the same powerplants and armament, but configured as reconnaissance platform with additional fuel in place of the bomb load.

He 343 D

Reconnaissance version similar to the He 343 C, but with Heinkel He S 011 turbine engines offering 1,300 kp static thrust each.

It is difficult to draw any conclusions about the He 343. Despite the different performance classes and the widely different roles for the Heinkel He 343 and Junkers 287 types, it is almost impossible to consider one type without the other. The He 343 was a desperate attempt to produce a multi-role aircraft for frontline service in light of the war's pressing question of whether or not it could be afforded. It was no longer a matter of whether it was needed, but rather, which was needed the more. A question that, with the imminent end of the war, remained forever unanswered.

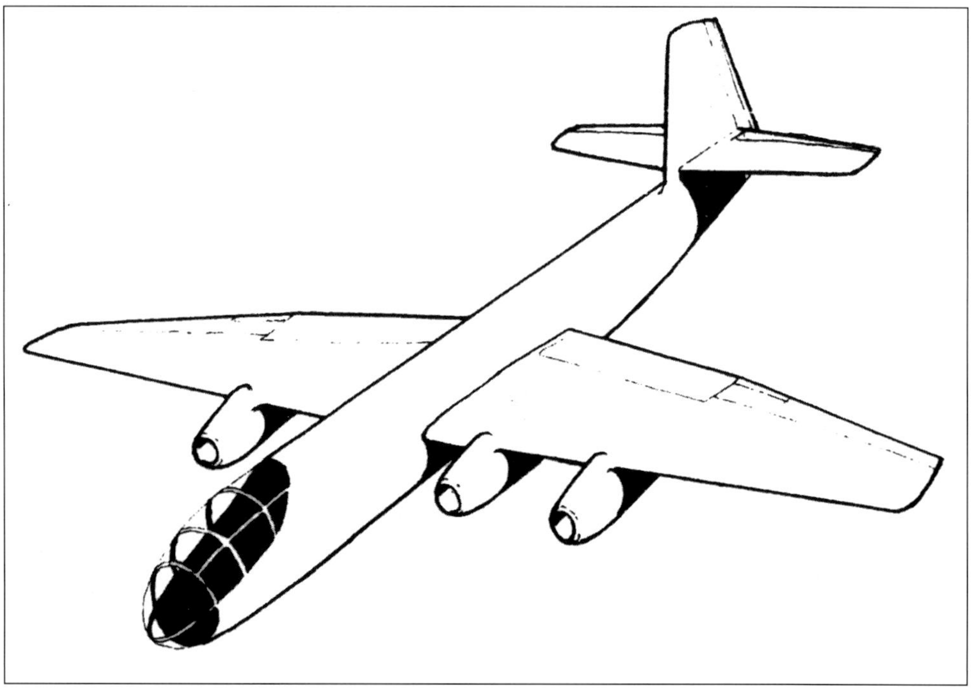

In conclusion, yet another fictional drawing of the Heinkel He 343.

Swept-wing Projects for Supersonic Flight

As was pointed out back at the beginning of this book, Germany was the only nation that both recognized the pressing need of swept wings for high speed flight and carried out research focused on this area. To be sure, after the war the Americans demonstrated that the sound barrier could be broken using a straight wing design, but this only occurred with a relatively high cost in power and at high altitudes where the air wasn't so dense. The more elegant approach to solving the problem undoubtedly lay with the swept wing, and initially it seemed insignificant whether a positive or negative sweep was employed.

To continue with the theme of swept wings and supersonic flight, we've included the box cover artwork for the DFS 346 1/72nd scale model kit from HUMA, which shows the supersonic fighter version - apparently a not too outlandish variant.

Four radically different projects were realized before the end of the war, and in one case there were even preliminary experimental flights carried out with a glider version.

The DFS tried it conventionally with the DFS 346, which had a wing with a 45% sweep and, using rocket propulsion, was to have accelerated to over 2,200 km/h at an altitude of 30,000 m. However, the DFS 346 was first and foremost envisioned as a test bed for supersonic flight studies. Since the history of the DFS 346 has already been covered in detail in my book "*Vom Höhenaufklärer bis zum Raumgleiter - Geheimprojekte der DFS*", I've quoted extensively from that title as there seems to be little more to add.

The Brothers Walter and Reimar Horten chose an entirely different path; they preferred a flying wing design with an extreme sweepback of 60%. However, for reasons which will be discussed later a central stabilizer fin had to be used, giving the flying wing a resemblance to the supersonic delta project of Professor Lippisch. The supersonic flying wing, with the designation Ho XIII b (Ho X) was to have been powered by a standard production turbine engine and go into production as a supersonic fighter.

As has already been alluded to, Professor Lippisch also planned a supersonic fighter. This was to have been conceived as a delta design and was given the designation of Li P 13. In order to study the behavior of the delta fighter concept, a glider version with the designation of DM 1 was first constructed. The Li P 13, unlike the other two previously named projects, was not to have been powered by either a rocket or a turbine; rather, Professor Lippisch favored a special pulse-jet engine that will be looked at later.

The most cautious, or even better - the most circumspect - approach to the problem was that of the Messerschmitt Works. Since there was no agreement as to which sweep angle was the most appropriate, the Me P 1101 V1 was created as an experimental design capable of conducting test flights using a wing that could be set at different sweep angles on the ground. In this case power was provided by a production turbine engine. The research data thus gained was to then have been incorporated into a later production version.

As can be seen, these were four entirely different concepts, four completely different paths, all leading to the same goal - a controllable, safe flight in the transonic region as well as an equally safe penetration of the sound barrier and subsequent supersonic flight.

DFS 346

The development of the DFS 346 has already been addressed earlier on page 12 onward, so that we can now take a closer look at system testing of the type. In addition to the Siebel Works' design team and technical experts, chief test pilot *Flugkapitän* Wolfgang Ziese was also deported to the Soviet Union, as had happened to those involved with the EF 131. In preparation for controlling the DFS 346 from the prone position the company had constructed a specially designed glider, the so-called "*Liegekranich*", as well as a "*Liegebaby*" (lit. "Prone Crane" and "Prone Baby", respectively). The "*Liegebaby*" was the primary platform used to train pilots on "Lying Flying". It was with this modified sailplane that Wolfgang Ziese and his Russian colleagues P. I. Kasmin, Mark Gallai, and N. S. Rybko familiarized themselves with operating an aircraft while lying prone.

At this point it is worth mentioning a rather interesting commentary based on recent information pertaining to the history of the DFS 346. Up to this point the DFS 346 was a pure experimental aircraft, but apparently there were plans being drawn up to make

The *Liegekranich* shortly after landing, with spoilers still deployed but sitting on its trolley ready for the next flight. The student lay prone in the forward cabin, while the instructor sat in the normal position in the second cockpit on the fuselage spine.

Another picture of the *Liegekranich* offers an overall impression of the aircraft shortly after landing.

it into something more. Around the end of the war a group of pilots were trained at Langensalza (Thuringia) on the "*Liegekranich*" and the "*Liegebaby*" in preparation for flights with the DFS 346. One of these was Wolfgang Ziese, but he was not alone. The so-called "Group 346" most likely consisted of ten of these pilots. Mr. Günter Gauglitz belonged to an anti-aircraft unit that at one time was stationed in Langensalza to protect "Group 346". This circumstance save Mr. Gauglitz's life, since, shortly before the Russians overran Langensalza, "Group 346" was pulled out of the area together with its flak unit. The units which stayed behind suffered a terrible fate. The Russians took no prisoners; everyone who had survived combat was summarily executed. The training involving such a large number of pilots for the DFS 346 raises a few questions. Was there perhaps a desire, assuming the trials were successful, to turn the experimental plane into an operational fighter after all? Was there already possibly a concept study for conversion, and they wanted

Another type used for training pilots in the prone position was the *Liegebaby*. It is evident that there was much less room available than in the *Liegekranich*. In addition, there was no second cockpit for the pilot instructor - the student was completely on his own, not counting the fact that he could receive instruction via radio.

to have the first batch of pilots ready beforehand? For now, these questions are purely hypothetical in nature - there is no evidence whatsoever for such speculation, and the interested reader is once again offered an opportunity to shed light on this matter. But even if such plans were being tossed around, the DFS 346 would most certainly not have been very successful as a fighter. Other than being able to fly quite fast in level, straight flight, it had no other qualities that would make it predisposed as a fighter. Similar to the F-104 Starfighter - an unsuccessful design for the same reasons appearing at the end of the Korean War - it would simply not have been in a position to outmaneuver a potential adversary. But enough of speculation; let us now turn back to real matters in the history of the DFS 346.

The next step in experimentation was the 346 P, towed by a cable behind a twin-engine B-25 Mitchell down the long runway at Tyopli Stan. This airframe only carried out small hops with subsequent landings on a belly skid. For these trials a special three-wheeled trolley cart had to be built, upon which the DFS 346 with its skid already extended would rest. A parachute deployed at the moment of lift off, bringing the trolley to an immediate stop similar to the takeoff trolleys for the initial prototypes of the Arado Ar 234; this was aided by automatically activated wheel brakes. At an altitude of a few meters the pilot Ziese, who carried out all test flights in the 346 P, would uncouple the 346 P and then land on the sprung skid.

Parallel to skid testing the rescue system on the 346 was also evaluated. A pressurized cockpit was carried, also by a B-25, up to an altitude of 9,000 meters and dropped. The system functioned reliably. At an altitude of 4,000 meters the pilot's seat separated on glide rails from the pressurized cockpit, which was rapidly plummeting to earth stabilized only by a small parachute, and the mannequin sitting comfortably on the seat then drifted down safe and well beneath another parachute.

Now the next stage of testing could be implemented, the testing of the 346 P in unpowered free flight. The B-25 was not suited for carrying the aircraft to altitudes above 10,000 meters, though, and it was decided to carry the 346 P to altitude beneath the starboard wing of a four-engine Boeing B-29.

Many a reader might incredulously rub his eyes and quite rightly ask: "B-25, B-29 - these are all American types. How did they get to Russia?"

Now, Russia and America were allied nations during the Second World War, and among other things Russia also received several American B-25s as part of aid packages. In addition, during the war against Japan about three of the four-engine B-29s made emergency landings in Russia. The Russians got them back into flying condition and then, after the war, it was easier for the Russians to simply "forget" to return them. Stalin was so impressed by this aircraft after the first atomic bombs were dropped on Hiroshima and Nagasaki that he ordered the B-29 to be copied. Thus, as the Tu-4, the B-29 enjoyed a sort of Russian Renaissance.

For the 346 P a type of "made in Germany" decoupling mechanism was used that proved to be quite reliable. The 346 P was released at an altitude of about 10,000 meters and Ziese was able to fly his test routine and subsequently touch down safe and sound on the skid of his 346 P at Tyopli Stan. During this program, his steep dives saw the 900 km/h mark exceeded several times. Wolfgang Ziese made about 30 such test flights, all of them without incident.

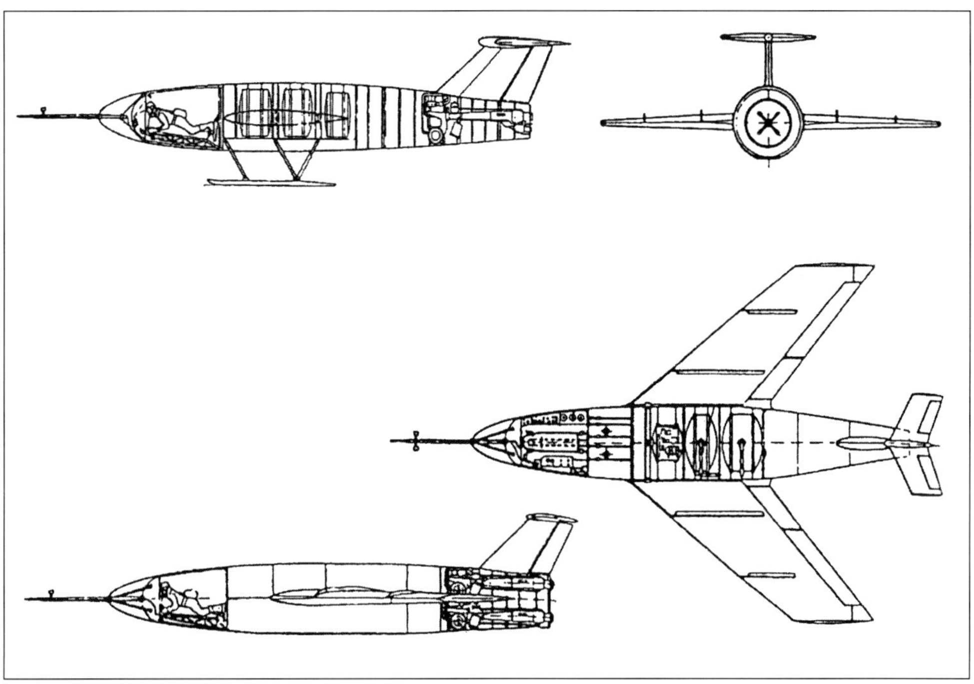

Three-view of the DFS 346 with side view of versions with one and two Walter rocket engines.

Close-up of the aerodynamically streamlined canopy glazing on the DFS 346 P non-powered glider version.

Entry into the cockpit was gained by sliding the entire glazing from the fuselage itself forward, to include the prone bench for the pilots.

Test pilot Wolfgang Ziese is lying well buckled in the narrow cockpit. The glazing can now be closed and hermetically sealed with the fuselage.

The DFS 346 P with Wolfgang Ziese on board, suspended beneath the wing of the mother ship - a Russian B-29 (Tu-4) - ready for takeoff.

On 5 May 1949 construction of the first 346, intended for rocket powered flight testing, was finally completed. All in all there were four prototypes laid down for these powered test flights. Preparations for the first flight dragged on throughout the summer until well into September 1949. On 30 September Wolfgang Ziese was finally able to climb into the 346 I, as it was now known, and make the first flight. The initial flights were to be carried out without power, with only a dummy engine installed and ballast compensating for the loss in weight. Takeoff weight of a 346 I configured like this was 3,145 kg.

The 346 I was hooked up to the drop mechanism underneath the B-29's starboard wing in the usual manner. As he had done so often before, Ziese squeezed himself into the narrow confines of the cockpit tube. Then the team took off. The B-29 now had markedly more weight to lift than before and climbed at a shallow angle to altitude. At 9,700 meters Ziese released. The glide back to earth revealed some unexpected problems that were thought to have been eliminated beforehand. In some flight attitudes the 346 I was even uncontrollable, but Ziese always succeeded in regaining control of the machine again. To be sure, similar problems had cropped up with the pure glider version, but not in nearly as critical a manner, so that it was felt the whole thing wasn't that important and worth continuing flight testing. Yet now, with the machine becoming heavier and the wing loading increasing, control quickly became unexpectedly more difficult. The problems during the gliding descent to the airfield precluded Ziese from setting up a proper landing approach. He came in at a much too high speed, at around 310 km/h. Although he touched down at the landing point, the machine jumped three or four meters back into the air and then, carried by the ground effect, continued along for another 700 to 800 meters before the machine finally touched down again with a hefty bump (certainly an unpleasant memory for every flight student who daily lands at too fast a speed. Monday - bump, Tuesday - bump, Wednesday- ... I don't think I need to continue). The solid landing caused

The attached DFS 346 P once again, seen from another angle. Clearly evident are the capped off openings for the rocket engines, which was a main identifying feature of the engine-less glider version.

The mother ship with attached DFS 346 P just after takeoff. The photo shows the large payload the B-29 needed to lift into the air.

the skid to pop out of its locked position and be forced back into its bay, so that the aircraft skidded along just like it was making a belly landing. The sudden deceleration prevented it from bouncing into the air again, but the effect was so great that Ziese's restraining belt could not hold his body back and tore loose. Ziese's body shot forward and he struck his face on one of the canopy braces. The impact was so great that it caused him to lose consciousness. This is how the ground crew found him, covered in blood lying in the shattered cockpit glazing. Fortunately, his wounds were not as serious as initially feared, and after a short stay in the hospital Wolfgang Ziese was able to fly again.

In the meantime, the Russians were taking their time again. A special investigation commission had been set up under the direction of test pilot N. S. Rybka, with the goal of identifying the circumstances that led to the accident. This commission came to the conclusion that the landing skid was apparently not fully extended; although able to withstand the first touchdown impact, on the second, more powerful one it collapsed back into the retracted position. It was speculated that the reason the landing skid wasn't fully extended in the first place was because Ziese had not moved the lever extending the skid to the fully locked position. It was assumed that Ziese, badly shaken by the events of the flight, had focused his full attention on maneuvering the aircraft and missed the fact that the extension mechanism lever was not fully locked.

The damaged machine was immediately repaired, but further flight testing was carried out by Russian pilot Pyotr Ivanovich Kasmin. Apparently trust in Wolfgang Ziese had been seriously damaged by the accident; it smelled of sabotage.

But as they say, the devil's in the doing. On his first test flight with the repaired 346 I the same misfortune befell P. I. Kasmin as had Ziese. Again the skid did not correctly deploy. Yet this time the landing occurred on snow covered ground, so that damage to

Despite the grainy quality of the picture, this photo was chosen for publication because so few exist of the DFS 346 in flight. The picture shows the DFS 346 on approach. Of note is the extremely shallow approach angle of the DFS 346 - highly unusual for such a fast aircraft.

The 346 I after its crash landing, which saw Wolfgang Ziese suffer a bloody nose.

man and machine was quite minimal. P. I. Kasmin suffered another accident on his second flight in the 346 I. Since his experience with the 346 was very limited, he miscalculated the distance back to the runway and was unable to make the airfield, being forced to land outside the base. Once again, the aircraft was broken and had to be sent to the repair shop for the third time.

In the meantime the ground tests with the 346's rocket engines had proven successful, and there was now a push for powered flight in the near future. Ziese, meanwhile, had regained the trust of his superiors following the accident of his unfortunate Russian colleague, and was once again permitted to carry out test flights. These flights, however, were to take place in Luckovice, about 120 km southeast of Moscow, meaning that a move to inhospitable environs had to take place first. During this interval the 346 II and 346 III prototypes were also finished. As there are no documents showing that the 346 II was ever flown, it is assumed that this machine was used exclusively for static stress experiments and wind tunnel tests at ZAGI.

But let's return to test flight operations. From October 1950 onward the test flights resumed and continued for about a year. In the meantime a captured and still flyable Ju 388 was used as a tow plane as well as a mother ship, alongside the trusty B-29 (which was later traded for its Russian counterpart). Prior to these flights, however, further test flights were carried out with the pure glider version, the 346 P, ostensibly for familiarization. The unpleasant problems associated with the 346 I's flight testing mentioned previously led to the decision to limit the maximum speed for the 346 I to 0.8 Mach and 0.9 Mach for the 346 III. Penetration of the sound barrier was considered critical, but there was disagreement on whether it should be risked. The higher authorized limiter for the 346 III was due to the fact that its wing configuration differed from that of the 346 I. The 346 III made use of a different wing profile.

On 10 May 1951 Wolfgang Ziese took off on an unpowered test flight with the 346 I. On 16 June 1951 there was another, also unpowered, flight with the 346 III. These test flights with the two

Another photo of the crash-landed 346 I.. Damage to the bent doors for the landing skid and in the lower fuselage area is evident.

Detail photo of the DFS 346's landing skid, the cause of many a problem on landing.

machines occurred without incident, and there was now a serious effort to transition to powered flight using rocket propulsion. For safety reasons, only a single rocket engine was to be ignited initially.

The first flight using rocket power took place on 15 August 1951, in the 346 III. At an altitude of 7,000 meters Ziese released himself from the mother ship. Approximately one minute and forty seconds later he lit the rocket engine and let it burn for ninety seconds. The machine accelerated forward rapidly, bucked upward slightly and then shot heavenward. During the climb, which went relatively smoothly, the familiar control problems cropped up again depending on the angle of attack. In addition, there was a certain instability when turning that could be offset with aileron application, however. After completion of this first powered flight Ziese once again landed at Luckovice without problem.

Barely recognizable from the poor quality of the photograph, this is reputedly a DFS 346 during a high speed flight at about 10,000 meters. If this photo is real, the small "bump" on the forward fuselage is something of a puzzle. Possibly a retro-fitted sensor?

The second flight under rocket propulsion took place on 2 September 1951 and was carried out without any incidence whatsoever. During the test flight the 900 km/h barrier was crossed for the first time in a shallow climb without any obvious problems occurring.

During his time in the Soviet Union Wolfgang Ziese was in constant correspondence with relatives in what was then West Germany. Naturally he was unable and not permitted to mention anything about his top secret activities in the Soviet Union; instead he had to do this cryptically. Such messages went something like this: Went for a bicycle ride today. I fell off and got my face banged up... This was as much to say: I carried out a test flight today. There was a crash landing and I was injured in my face. I'm okay...

On 14 September 1951 the third rocket-powered test flight took place, this time with the 346 III. Ziese filed the following report on this flight:

> I released the machine at an altitude of 9,300 meters and made a smooth descent to about 8,500 meters. As I'd done twice before, I ignited the engine, but this time the airplane immediately began to climb rapidly. Although the flight speed remained constant at first, it suddenly jumped higher. At 12,000 meters' altitude I had long since crossed the 900 km/h barrier, yet the machine continued to accelerate. Suddenly, as had so often happened before, the machine no longer responded to control input. The 346 III immediately bunted over into a dive, and I thought I was a goner, but at about 8,000 meters I was able to regain control of the aircraft. I wanted to breathe a little bit easier, but at 7,000 meters I lost control again. There was no way I was going to get this machine back on the ground

Another head-on shot of the B-29 with an underslung DFS 346 P concludes the chapter.

> like this, so I decided to eject the cockpit capsule at an altitude of about 6,500 meters. I fell inside the capsule to about 3,000 meters, then I was suddenly extracted from the capsule and found myself floating freely. There was a jerk, and I hung beneath my parachute. Thank God that the rescue system worked.

Wolfgang Ziese made it back to Mother Earth safe and sound. Several other publications have in the past have reported that the 346 lost a wing during supersonic flight at an altitude of 20,000 meters and that Ziese broke both legs when he made his parachute landing. Others have reported that Ziese reputedly died during the single rocket-powered flight. All these reports share nothing in common with actual events. A wing break at transonic speeds or greater would have caused the 346 III to subsequently go into such a powerful roll, with unimaginably high centrifugal forces, that the whole machine would have broken up almost immediately. Therefore, it is completely inconceivable that Ziese would have then been able to save himself; the centrifugal forces alone would probably have killed him outright. Since Wolfgang Ziese died from cancer on 28 August 1953 at the relatively young age of just 46 years, he could not therefore have already died during a test flight in 1951. Wolfgang Ziese was buried at the cemetery in Podberesye.

But back to the events of 14 September 1951. The 346 III was totally destroyed upon impact with the ground, and the recording equipment was damaged so much that it was impossible to assess the flight. Nor could the cause of the crash be clarified without much difficulty. There was still a bit of uneasiness about the incident mentioned earlier, however. The rumor was that Wolfgang Ziese was feeling his oats, so to say. He couldn't wait and fired up both engines in order to demonstrate that after the previous problem-free flights the sound barrier could finally be 'broken'. He was tired of the plodding test flights by the Russians and wanted to bring the flight program to a rapid conclusion. Fortunately for him the cause could not be determined. If he really did attempt to punch through the sound barrier at full throttle, then at least he did bring about a quick end to the program, for testing came to a halt after the loss of 346 III. 346 I was not suited to breaking the barrier, and no other machines were to be built. In 1953 the OKB-2 developmental bureau at Podberesye was disbanded and the workers were sent back to Germany together with their families. Thus ended in a particularly unspectacular fashion the foray into supersonic flight.

The test flights provided valuable information for designing modern supersonic aircraft, and the data is still relevant today. The next chapter will present a swept-wing project that, without any equivocation, can truly be considered 'fantastic' - the Horten Ho XIII.

Horten Ho XIII (Ho X)

Long before the SS could stretch their feelers out toward the developments of the Brothers Horten, these men had drawn up a design for a supersonic flying wing with the unbelievable sweepback of 60%. Never before had an aircraft flown with such an extreme wing sweep, particularly not a flying wing design, unusual enough on its own. In a transferred sense, this applied equally to the DFS 346 described earlier, for up to this point there had been absolutely no experience with extreme sweepback.

Work on the configuration of this daring project began quite modestly in 1943. Sometime in March 1943 work transferred to the *Autobahnstraßenmeisterei Göttingen*. Planning work took place in two newly erected barracks. For construction of the Ho XIII the spacious accommodations of the *Autobahnmeisterei Hersfeld* was monopolized, which as a branch was also responsible for high speed research.

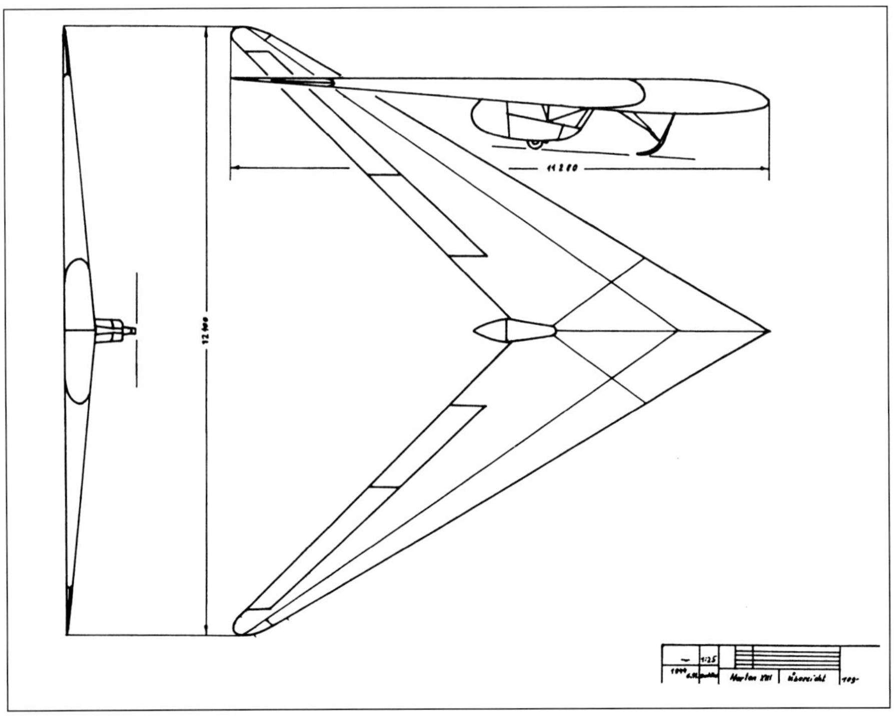

A simplified three-view drawing of the Horten Ho XIII reconstructed from the original dated 11/6/1944.

The Horten Ho XIII a, with fixed landing skid, parked in front of the hangar at the airfield in Göttingen. This photo shows the unusual side silhouette of this flying wing.

The Ho XIII a from behind, taken at the same hangar. (Photo: H. P. Dabrowski)

The right wing has been removed, allowing a clear view of the cockpit gondola and the massive supports for the landing skid. (Photo: H. P. Dabrowski)

In order to rapidly obtain data, particularly with regard to slow-speed flight handling of the extremely swept flying wing design, a powered glider version was initially built (as was the case with many other new projects). The wings of a Ho III b were needed, though these were modified so that there was a 60% sweep angle at the wing/fuselage joint. Only the fuselage center section required total redesigning. The new center section effectively consisted of a type of gondola, which hung beneath the 12 meter span of the wings and had wing attachment points on either side. The pilot could enter this gondola by opening a hatch at the rear. In an emergency, he could just as easily and quickly escape by parachute through the selfsame hatch. However, for the pilot this accessibility came at an enormous price; pilot visibility was seriously restricted, in a word, it was absolutely miserable. This would later lead to this or that critical problem during testing, as we shall soon see.

The Ho XIII designation was undoubtedly influenced by the fact that the wings came from an Ho III b; in addition, the supersonic project was in fact carried under the designation of Ho X, a follow-on project to the Ho IX that flew as a twin jet aircraft in the subsonic region. Out of necessity, the two Roman designations from the Ho X and Ho III (the wing donor) were combined, giving the equation X+III = Ho XIII, the designation for the test glider. Reimar Horten claims in statements made after the war that this designation was also a cover to ensure secrecy. Be that as it may, the Ho XIII designation stuck and a simple letter suffix distinguished the glider version (Ho XIII a) from the later, powered version (Ho XIII b).

Hermann Strebel was assigned responsibility for testing the Ho XIII a once this glider version was finished in November 1944. For its trials the machine was brought to the Hornberg branch (Göttingen). Above all, the Ho XIII a was a high performance glider, with a glide ratio of 12, i.e. for the uninitiated, this means that the Ho XIII could, from an altitude of one kilometer above ground with no wind, only glide a distance of 12 kilometers (compare this to a modern high performance glider that can easily cover a distance of 35 kilometers and more, giving it a glide ratio of 35 or greater).

On 17 November 1944 an initial attempt at flight was made by towing the Ho XIII a behind a car, and indeed, the Ho XIII managed to make a little hop into its element. On 27 November 1944, a day as peaceful as could be with just a hint of a breeze at 2 meters per second, the first flight took place as the glider became airborne towed by a Henschel Hs 126. Fortunately, Hermann Strebel's flight report has survived, and he states:

> Since the hop on 17 November 1944 the center of gravity and takeoff weight had remained unchanged. The aircraft took off against a westerly wind of about 2 m/s from the southeastern corner of the Göttingen field to the northwest. The tow plane was an Hs 126 with *Oberfeldwebel* Knöpfle as the pilot.
>
> Takeoff was normal, with tow speed hovering between 130 and 140 km/h. At that speed aileron and elevator effectiveness were adequate, but rudder control was unsatisfactory, and the negative yaw moment was low.

A very good detail photo of the cockpit gondola. One can see that the forward view is somewhat "obstructed". The jettisonable end cap is also quite prominent. (Photo: H. P. Dabrowski)

Above: Aerial view of the Ho XIII a taken from above reveals the extreme sweep of the wings to good effect. The Ho XIII a looks like a flying arrowhead. Right: Another aerial photo yet again reinforces the impression of a flying "arrowhead" (Photos: F. P. Selinger)

The unusual shape of the Ho XIII a from the side shows another view of the cockpit gondola under the wing. (Photo: F. P. Selinger)

The view forward is poor, but is adequate for towed flight. I uncoupled at an altitude of 2,700 meters and flew straight ahead at speeds between 70 and 150 km/h. Aileron effect at normal speeds is only slightly less than that of the Ho III, but drops off as speed increases and at 150 km/h it is so low that a bank cannot be maintained. Elevator effect is adequate at all speeds, although the travel is somewhat too little. With elevators full up, speed is 70 km/h, and at this speed no airflow separation is apparent. Rudder effectiveness is quite limited, but it does allow the aircraft to turn without problem. Turning with a bank angle of about 45% there were slight oscillations around the roll axis, with their amplitude remaining below 5%; these could not be fully suppressed with counter-application of ailerons, but could be minimized considerably.

Landing took place without problem. When gliding in on approach the poor forward visibility became notably troublesome. The point of touchdown cannot be seen until leveling out, something that would make landing at an unfamiliar field difficult. Because the aircraft could not be fully leveled, it bounced a few times after touchdown and then rolled in a straight line. On the third bounce the machine was still traveling at a speed of 50 km/h. The wheel brakes did not cause the wheel to lock up, but reduced speed quickly enough. There can be no conclusions drawn about the flight handling of the Ho XIII a; with fully deployed flaps the aircraft continues to glide along for some distance when at ground level.

A further 18 to 20 experimental flights were carried out before the war's end. During one of these flights there occurred a long overdue incident on landing. The poor visibility caused the Ho XIII a to drift too far along the runway and touch down much too late. At a good clip it ran into a perimeter fence at the edge of the field. The impact speed was so great that the fence posts broke in two, as can be seen in the photo on page 173. The pilot was uninjured. Damage to the aircraft and the fence was so minimal that both could be repaired in relatively short order. In addition to the Hs 126 mentioned previously, another tow plane was the Focke-Wulf Fw 58 *Weihe*, and it was this type that was used on 18 December 1944. The unpowered free flight began after a towed flight of 26 minutes, lasting ten minutes without any particular problems. This time the landing went smoothly.

The trials continued at a relaxed pace without any real problems, when the SS assumed control of all advanced design developments that offered promise. It wasn't until relatively late in the game that they got wind of the supersonic project.

The picture changed in the spring of 1945; there was considerable interest in fielding a superior supersonic fighter as quickly as

The Ho XIII a in a left bank on landing approach reveals the ever present landing skid. (Photo: F. P. Selinger)

In the foreground is a model variant of the Ho XIII (Ho X), with Reimar Horten standing behind wearing the peaked cap. (Photo: F. P. Selinger)

Data table for the Horten Ho XIII a.

Role			**experimental**
Construction	**Fuselage (center section)**		**steel tubing**
	Wings		**wood**
Crew			**pilot**
Wingspan		**m**	**12**
Leading edge sweep angle		**degree**	**60**
Taper ratio		**6.5**	
Root thickness		**%**	**12**
Root chord depth		**m**	**5.2**
Rib spacing		**m**	**0.4**
Wing area		**m^2**	**36.0**
Aspect ratio		**4.0**	
Pilot position			**seated**
Width, center section		**m**	**2.4**
Width, cockpit		**m**	**0.6**
Height, cockpit		**m**	**1.0**
Weight, empty		**kg**	**250**
Ballast (e.g. water)		**kg**	**-**
Load (payload)		**kg**	**60**
Max. flight weight		**kg**	**330**
Wing loading		**kg/m^2**	**9.2**
Stall speed		**km/h**	**44**
Landing speed		**km/h**	**44**
Minimum sink rate		**m/sec**	**1.1**
		at km/h	**60**
		and kg/m^2	**9.2**
Best glide		**16**	
		at km/h	**80**
		and kg/m^2	**9.2**
Not to exceed speed		**km/h**	**180**

possible. The production capacity of the Brothers Horten was stretched nearly to the limit with the simultaneous running of their two projects, the twin-engine Ho IX flying wing jet and the development of the six-engine Ho XVIII turbojet strategic bomber.

Nevertheless, they were able to produce several designs for the Ho XIII b, as the supersonic jet fighter project was now designated. In order to ensure a stable attitude during high speed flight it was critical to fit a central rudder. The background behind this necessity was revealed in the tests with the Ho XIII a; with a half-span of just 3.6 meters and a length of 10 meters drag rudders were no longer sufficient. Furthermore, drag rudders would cause compressibility problems on the wing upper surfaces when penetrating the sound barrier.

Brief explanation: How does one fly a flying wing? Turns made with so-called "drag rudders". This is understood to mean extendable tabs fitted to the ends of the wings that generate moment through the drag they induce. By an asymmetrical extension of the drag rudder the flying wing is pulled to one side, and supported by conventional ailerons a coordinated bank can be initiated.

In the end, preference was given to two designs (see drawings p 172), one with the pilot housed in the rudder base (remarkably similar to the Lippisch Li P 13a design, see page 192), or another with the pilot in a prone configuration in the upper forward fuselage. Only a few successful model experiments could be carried out before the end of the war. Powered by a Heinkel He S 011 A-0 turbojet engine, speeds of 1,200 km/h were expected at an altitude of 6,000 meters, which equated to a Mach number of 1.07. Supposedly, a wooden glider version of the Ho XIII b was under construction around the time the war ended that was captured by the British. There was reputedly even the idea of fitting the machine with the Argus As 10 c engine, an Argus ramjet, prior to actually testing it with the Heinkel turbine. Perhaps a reader might know more about this and can offer some information.

Just after his successful maiden flight, pilot Hermann Strebel (right) briefs Rudi Opitz about his impressions and experience. (Photo: K. Nickel)

The Ho XIII a after its landing accident. One can clearly see the bent fence post on the right edge of the photo. To the left, under the wing, is the separated fuselage end cap, enabling the pilot to exit the cockpit gondola out the back. (Photo: F. P. Selinger)

Crew:
pilot in reclined position in a cockpit blended into the wing outline, protected by curved plexiglass canopy

Propulsion:
1x Heinkel He S 011 A-0 with 1300 kp static thrust, later more powerful version with 1500 kp static thrust; additional booster rocket planned

Dimensions:

Wingspan:	**7200**	**mm, delta wing with 70% leading edge sweep**
Wing area:	**37.8**	**m², aspect ratio 1.37**
Profile thickness:	**7**	**%**
Total length:	**10000**	**mm**
Max. height:	**ca. 4200**	**mm**

Weights

Takeoff:	**6000-7000**	**kg**
Wing loading on takeoff:	**160-185**	**kg/m²**

Performance

Max. speed:	**1200**	**km/h (Mach 1.07) at 6 km altitude**
Service ceiling:	**15000**	**km**
Range:	**2000**	**km**

Military equipment

Armament:	**3-4 MK 108, cal. 30 mm, or 3-4 MK 213 C, cal. 30 mm**

The Horten Ho XIII a, destroyed by slave laborers following their liberation.

Another photograph of the damaged aircraft with the unfortunate pilot, Hermann Strebel, standing alongside. The writing on the sign to the left proclaims in a purely mocking tone: "Taxiing aircraft outside the runway area is only permitted on hard surfaces"! A restriction that the Ho XIII a obviously ignored. (Photo: K. Nickel)

This photograph shows just how much the "free-spirited" Ho XIIIa dared to cross over the perimeter fence. (Photo: H. P. Dabrowski)

The remnants of the cockpit gondola on the destroyed Ho XIII a. (Photo: H. P. Dabrowski)

The repeatedly published portrayal of the Horten Ho III b (Ho X) as envisaged by Gert W. Heumann.

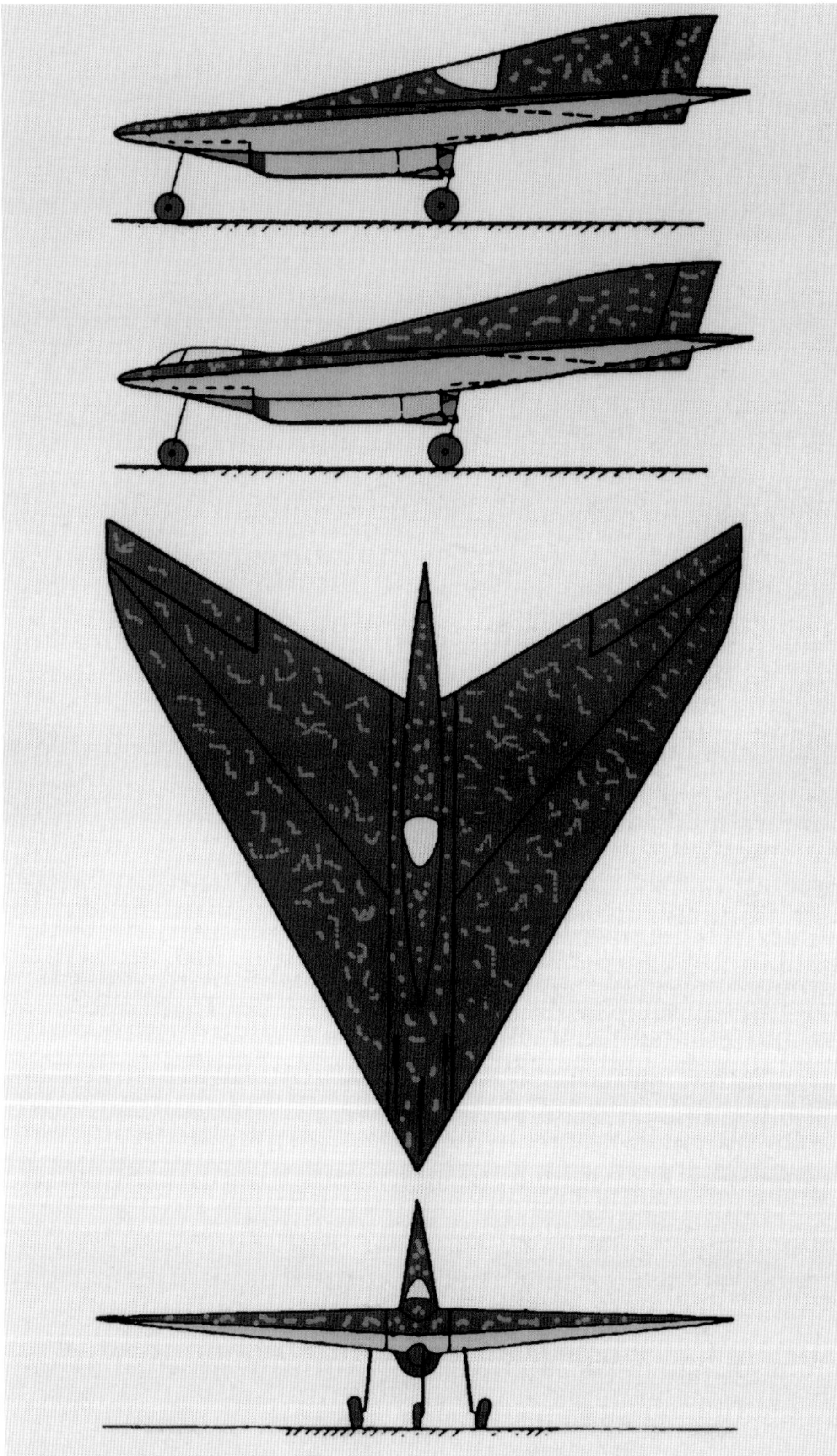

Three-view sketch of the two planned variants of the Horten Ho XIII B, as it probably looked based on statements by co-workers on the project. The above view shows the pilot housed in the rudder fin, where he would certainly have had excellent visibility, and an alternate version with the pilot lying prone in the forward fuselage section. The latter would have enabled the pilot to withstand greater *g* forces in dogfights and when pulling up sharply, permitting tighter turns at higher speeds.

Lippisch LiP 13 (*Triebflügel*) and DM 1 Experimental Glider

The Li P 13, a supersonic configuration of Professor Alexander Lippisch, combined several radically new conceptual characteristics, advancing the combination of engine, fuel and airframe in entirely hithertofore unseen dimensions. In order to highlight the problems encountered as a result, it is necessary to explore the developmental history from the beginnings of the Delta flyers up to the Li P 13 in somewhat greater detail, as well as the ramjet propulsion system inseparably connected with the Li P 13. For it was the merging of these two developments that made the unusual concept of the Li P 13 possible in the first place.

Alexander M. Lippisch, who later became professor, had joined the DFS back in the '30s. Lippisch was well known for his unparalleled pioneering efforts in the area of tailless aircraft. Around 1930 there were numerous experiments carried out with the goal of building a viable flying wing aircraft. All these early designs suffered from rather severe stability problems. Professor Lippisch was

Air-to-air photo of the two-seat DFS 39 (Delta IV c), an early example of a design using downturned wingtips.

This photo shows the Delta IV a, also a two-seater, with standard wingtips, which served simultaneously as rudders.

tasked with tackling these problems and coming up with a practical solution for solving them. Thus was born the famous Delta series. Delta I was created in 1932 with the generous support of the well known transoceanic flyer Hermann Köhl. This was followed by the Delta II equipped with a 20 hp engine. Finally, there was the Delta III in 1933, two examples of which were built at the Fieseler Works in Kassel under the designation of F 3. All three designs incorporated end plates affixed to the tips of the wings that acted as vertical stabilizers. These designs suffered from the same problems as their unsuccessful competitors. After much research, the designers became convinced that the better approach would be to have the rudders attached to the wingtips, but angled sharply downwards and of a wedge-shaped design. The first experimental aircraft incorporating this configuration was the DFS 39, also known as the Delta IV. The following flight trials bore out the correctness of this thesis, and so the DFS 39 became the quintessential forerunner of tailless aircraft. From the Delta series of test aircraft came the later designation of such types as delta-winged planes, a common term even today for this class of aircraft. Professor Lippisch chose this designation at the time because the triangular wing shape

A group photo of the tandem powered Delta IV "*Wespe*" (Wasp) sport plane with upward folding wings next to a conventional Fieseler Fi 5 R does well to illustrate the advanced "delta concept" planform.

The natural developmental progression of the delta concept led to the DFS 40.

generally resembles the delta of a river. The DFS 39 ultimately spawned a two-seat machine with an 85 hp engine in the nose, which made its successful maiden flight in 1935. This aircraft was also known as the Delta IV d and proved to be quite a successful sport plane.

A continuation of this development was the DFS 40. This was an expansion of the delta series and was also known as the Delta V. The wing configuration and control surfaces were virtually unchanged. The most glaring difference was the fully glazed pilot's cockpit in the forward nose section. Also different was the prone

The DFS 194 was the first successful step in mating a delta aircraft with a rocket engine.

position the pilot was now in. Although it was originally to have been powered by a 120 hp Argus engine, the DFS 40 was only flown and tested as a glider. A Heinkel He 46 served as the tow aircraft, pulling the DFS 40 to altitude. Once the craft was in a free glide, the prone position of the pilot in no way proved to be disadvantageous. Pilots Heini Dittmar and Rudolf Opitz carried out extensive trials with the type. However, in the summer of 1939 the machine crashed when it unintentionally went into a flat spin and the pilot, Rudolf Opitz, was unable to regain control. The DFS 40 impacted in a grove of trees, totally destroyed. Miraculously, Opitz only suffered minor injuries.

But Professor Lippisch wanted more, always searching for a way to go faster. Thus was born the design for a rocket powered experimental aircraft with the designation of Project X. For this, Professor Lippisch pulled out the wing geometry of the DFS 39 and DFS 40, since he assumed that he could achieve higher speeds most quickly with a tailless configuration and that a rocket propulsion engine could be fitted in the aft section of the fuselage without any problem. Only the empennage was designed entirely from scratch and now consisted of a central rudder in the tail. The wings were constructed at the DFS, while fuselage assembly went to the Heinkel Works. The powerplant utilized was the HWK I-203, a liquid-fueled rocket motor with 300 kg thrust produced by Walter in Kiel. While still working on the project design Lippisch and twelve of his colleagues had to relocate to the Messerschmitt AG for reasons of space. Group L's design work continued apace at Messerschmitt. The DFS 194, already under construction, was completed at the Messerschmitt Works in early 1940, but did not go into flight testing until the summer of 1941. Again, it was Heini Dittmar who carried out the flights, and he was able to achieve speeds of up to 550 km/h with this rather modestly powered engine. The DFS 194 was, in fact, the immediate predecessor to the well known Me 163 *Komet*, the first rocket powered interceptor to go into production and the only one of its kind in the world to ever be used operationally.

The Me 163 B *Komet* was, without any great effort, able to log speeds of up to 1,000 km/h, and many a pilot in the heat of combat experienced the same unpleasant sensations as had Heini Dittmar when he broke through the sound barrier.

Professor Lippisch's next focus was on breaking the sound barrier with a production model aircraft. In the meantime he had relocated to the *Luftfahrtforschungsanstalt* (Aeronautical Research Facility), or LFW, in Vienna, and 1943 drew to a close. With cries for a simple, superfast fighter becoming louder and louder throughout 1944, his Li P 13 supersonic project became all the more critical. Deviating from all previously discussed projects, his supersonic fighter was to have been a delta wing configuration powered by a special kind of ramjet engine. Not content to use just a ramjet,

For a time, the Me 163 "*Komet*" (Comet) was the final link in the chain of successful rocket- powered delta-winged aircraft.

Trials on the Lüneburger Heide using a test engine mounted to an Opel Blitz truck. The man on the left in the white lab coat in a rather anxious pose is Professor Eugen Sänger. In the background can be seen the chase car, which monitored the test run from behind.

the development of which was still in its early stages, the package would be sweetened by using a mixture of coal granules and heavy oil as fuel - a flying "coal furnace".

Before examining this unusual development, a word is in order regarding Professor Dr. Eugen Sänger, who had been working on a functional ramjet development as early as 1940 and was the only one to have tested it in flight, as well as filing a patent under the number DP 411/42 on 22 April 1939 for a lifting body with an airspeed of over Mach 5!!! However, it wasn't until June 1942 that the patent was actually issued.

Around the end of 1940 the "Trauen Flight Test Center" hosted a VIP visit; high ranking generals from the *Luftwaffe* graced the site with their presence. There were discussions about the problems with the air war. It was expected that the Allies would be able to attack with bomber aircraft and that the Germans would have nothing to counter them with. Options and alternatives were sought out that would enable them to meet such a threat.

Subsequent to this visit, Eugen Sänger mused on the concerns and came up with the idea of a ramjet powered fighter having exceptional speed. The ramjet was the simplest form of propulsion imaginable. The principle had been known for a long time and hearkened back to a patent by René Lorin from 1908, although up until now no one had carried out a practical application of the idea that such an engine could be utilized effectively for an aircraft. The propulsion system consisted of a simple tube that was accelerated rapidly through the air. In the forward section the air intake was constricted. In turn, this constriction caused the inflowing air to build up and, because of the dynamic pressure, compression occurs. In the combustion chamber located behind the air intake, fuel is injected and ignited. The hot combustion gases flow out the back of the tube at a high velocity and accelerate the lifting body forward.

However, the engine had some inherently bad drawbacks. It only worked with any useful and acceptable efficacy after a certain

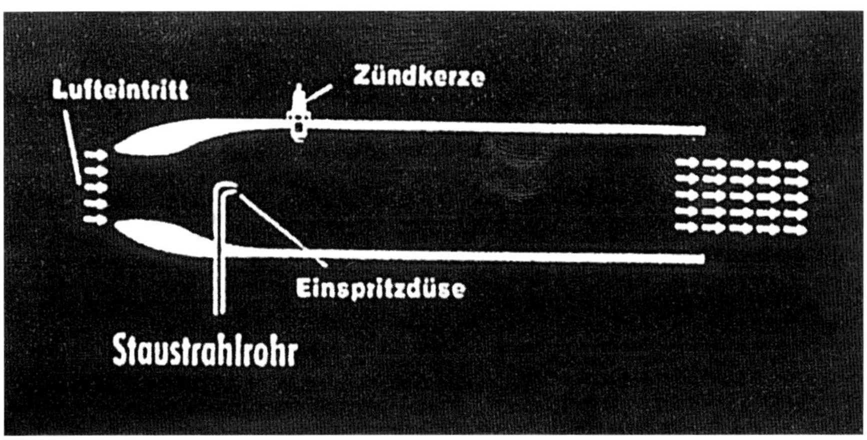

Concept diagram of a ramjet engine, a powerplant without any moving parts. The air is compressed through forward motion and forced into the engine, where it is mixed with atomized fuel and ignited. This produces thrust and offers a significant boost in airspeed.
Key:
Lufteintritt - air intake
Zündkerze - spark plug
Einspritzdüse - fuel injector
Staustrahlrohr - ramjet

speed had been reached. Another disadvantage was the enormously high fuel consumption. But Eugen Sänger's motto was, when problems cropped up, they were to be solved by staying focused on the problem, not chasing solutions that can take you away from the task at hand. Thus, he created the basis for the construction of the first ramjet engine, which was registered under patent number 165144 as an "ultra-high temperature ramjet propulsion system".

When the first experimental tube had been completed, one of the observers seemed to find the most fitting moniker for this type of engine, one that would be associated evermore with the type: "That looks just like an extra large stovepipe!".

For the first ground tests the engine was mounted on a rather aged Opel Blitz. The mounted test engine stuck out both ahead of and behind the truck. Even from the initial trials, it was apparent that the engine worked. With a lot of smoke and a huge tongue of flames, the truck moved forward - a real "hot seat", so to say. Different tube diameters were evaluated. However, no spectacular engine performance was expected at the low speeds with the Opel Blitz. After all, we've learned that the faster the forward movement the greater the effectiveness. What was needed was a flying test bed. The first test pilot for the project was *Flugkapitän* Paul Spremberg. Later, when flight trials were transferred to the DFS, it was Erich Klöckner as the pilot.

At the Faßberg airfield, the first experimental engine - with a diameter of 50 cm - was mounted to the back of a twin-engine Do 17 Z carrying the identification code of BC+NI. The separate twin rudder empennage made this machine ideal for the trials, as the exhaust jet had what amounted to a clear path. Paul Spremberg had mysteriously made this machine to appear "as if by magic". The first test flight took place on Friday, the 6th of March 1942. How would the Do 17 behave in flight with a giant object strapped onto its back? In order to avoid overloading the aircraft, it was flown with the fuel tanks only half full. On board were just *Flugkapitän* Spremberg and the 'mechanic' Eugen Sänger. Sänger did not want anyone else taking the risk, and so he himself flew along. The runway used at Faßberg was about 6,000 meters long, but railroad tracks crossed the middle of the runway. This unevenness would undoubtedly have caused the landing gear to collapse on the overloaded aircraft. It was with much effort that the Do 17 lifted off the runway after a run of about 1,000 meters and flew just a few centimeters above the ground. But Spremberg set the machine down again and wanted to roll out because the plane was not trimmed properly and was pulling to one side. "What's wrong?" called Sänger. "The machine's not keeping straight", came the short answer. "Doesn't matter, try it again!" called Sänger back. "Then we'll have to get airborne before we hit the tracks", responded

The small test engine, with a diameter of 50 cm, on the back of a Dornier Do 17 Z during a "hot" test run with the engine fired up.

The larger test engine, with a one-meter diameter, on the back of Dornier Do 217 RE+SP.

Spremberg. Just before the tracks the Do 17 came unstuck, climbed to about three meters and struggled to gain even more altitude. She flew.

They were in the air for about 30 minutes and flew every type of maneuver imaginable. The proof was there; the machine was fully maneuverable with the unusual object on its back. Real flight testing with the "flying stovepipe" powered on could now begin.

The first test flight with the engine lit took place on 7 March 1942 and offered a marked increase in speed for the Do 17. Spremberg felt a noticeable kick in acceleration when the engine was switched on at a speed of about 380 km/h. Level speed increased to 480 km/h until the limited fuel supply ran out. In high spirits, Eugen Sänger reported the success to the RLM in Berlin. But the response was indignant: "What? You're already working? You've not been given authorization for that!" Nevertheless, the trials continued. A larger and more powerful test ramjet, with a one-meter diameter, was mounted onto a higher performing Do 217 E-2. On the second flight with engine on the entire upper fuselage skin melted due to the intense heat. As a result, the Do 217 got a double upper fuselage skin, with a cooled gap of 5 mm between each layer, as well as a heat resistant coat of paint made from aluminum powder. In addition, the elevators were skinned with steel.

"Hot" test flight with the even larger ramjet engine having a diameter of 1.50 meters. The ramjet leaves a massive trail of smoke in its wake.

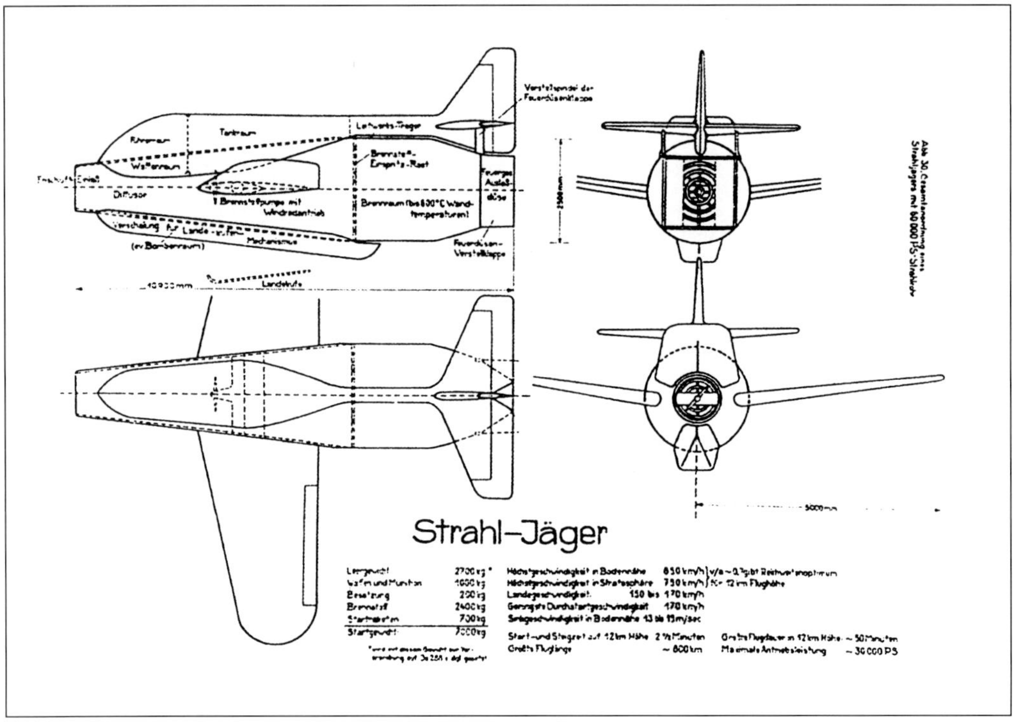

Early design drawing of a ramjet fighter as was to have been built by the Skoda Works. The engine was to have been powered by coal granules.

In the meantime, Berlin had been markedly increasing pressure on the program, and Sänger broke camp at Faßberg to move to the DFS at Hörsching. His friend and patron, Professor Georgii, had arranged for permission from the research leadership to continue the flight trials at Hörsching. Up to that point, through 12 June 1942 a total of 69 test flights had been carried out at Faßberg with the 50 cm tube, and a further 11 between 19 July and 08 September with the 1-meter diameter ramjet.

Meanwhile, a powerful test engine had been built that would be mounted on the back of the Do 217 E-2. This engine was to have the incredible and, for that time, unbelievable, output of 20,000 hp. Together with mechanic Kaffenberger, on 16 August 1943 Erich Klöckner flew the Do 217 from Ainring to Hörsching, where the machine would have this massive engine strapped onto its back.

After the initial test flight with the newly mounted engine, Klöckner noticed that the takeoff run had increased considerably and that the cruising speed was reduced from 440 km/h to a meager 320 km/h. The first 'hot' takeoff took place on 21 August 1943. Erich Klöckner still remembers this flight quite well, as well as his personal dealings with Eugen Sänger and Irene Bredt:

> During this first live test with the engine we climbed to an altitude of 3,000 meters and leveled out. Eugen Sänger sat to my right and below in the cockpit. The moment he lit the engine there was a rumbling noise, and you could clearly feel the pressure on your back. Speed increased from 320 km/h to an easy 400 km/h, but I felt an unpleasant vibration on the control stick. The dive angles increased over the next several test flights, and during these the airspeed increased to the maximum authorized for the Do 217, which was 720 km/h. The faster the airspeed, the more violent the engine's thrust became, and, however, also more unpleasant and harder the oscillations. Since the reason for these oscillations had, in the meantime, been discovered, they could have simply been stopped, but because of time constraints we lived with them and continued the testing program. During the first diving flight where we reached maximum authorized speed, on 4 October 1943, there was almost a catastrophe. After a normal test flight to that point and the unavoidable strong oscillations, I noticed on approaching the airfield at an altitude of 50 meters that I would be short and so gave it full throttle to extend the approach. The starboard engine rapidly took hold, but the port one remained docile. Suddenly we found ourselves coming in at an angle to the direction of flight, just as though we were making a slip. Only by applying hard right rudder could we get the nose pointed in the right direction

again. But at minimum speed at such a low altitude that would have inevitably resulted in a crash. There was nothing else to do but retract the landing gear, drop the flaps and try for a belly landing. Since the Do 217 had a fully glazed nose, when the lower panes shattered on impact a massive amount of dirt forced its way into the cockpit. After what seemed an eternity the machine finally came to rest, and there stood Eugen Sänger up to his knees in mud, unable to free himself. I, too, had trouble pulling my feet out of the soft earth. But the airfield fire team was on the scene quickly and together we freed Sänger from his precarious situation. Days later I learned that the fuel lines to the port engine had broken because of the powerful vibrations. Testing resumed in March of 1944. By 30 August 1944 I had carried out a total of 33 additional flights, but further flights were prohibited after that due to fuel shortages. Sänger then limited himself to wind tunnel tests. From time to time the research team of Eugen Sänger and Irene Bredt paid us a visit. Since we had a relationship with the owners of a farm, we always had plenty to eat. During a stroll, Eugen Sänger told me that the time will soon come when mankind will be able to fly into space with the aid of rockets and even land on the moon. I must confess that, at the time, I looked at him with something akin to horror on my face, thinking: "he's not crazy, is he?" He also mentioned an engine that would one day permit man to fly at the speed of light. He called it an ion engine. It was fascinating to hear him talk.

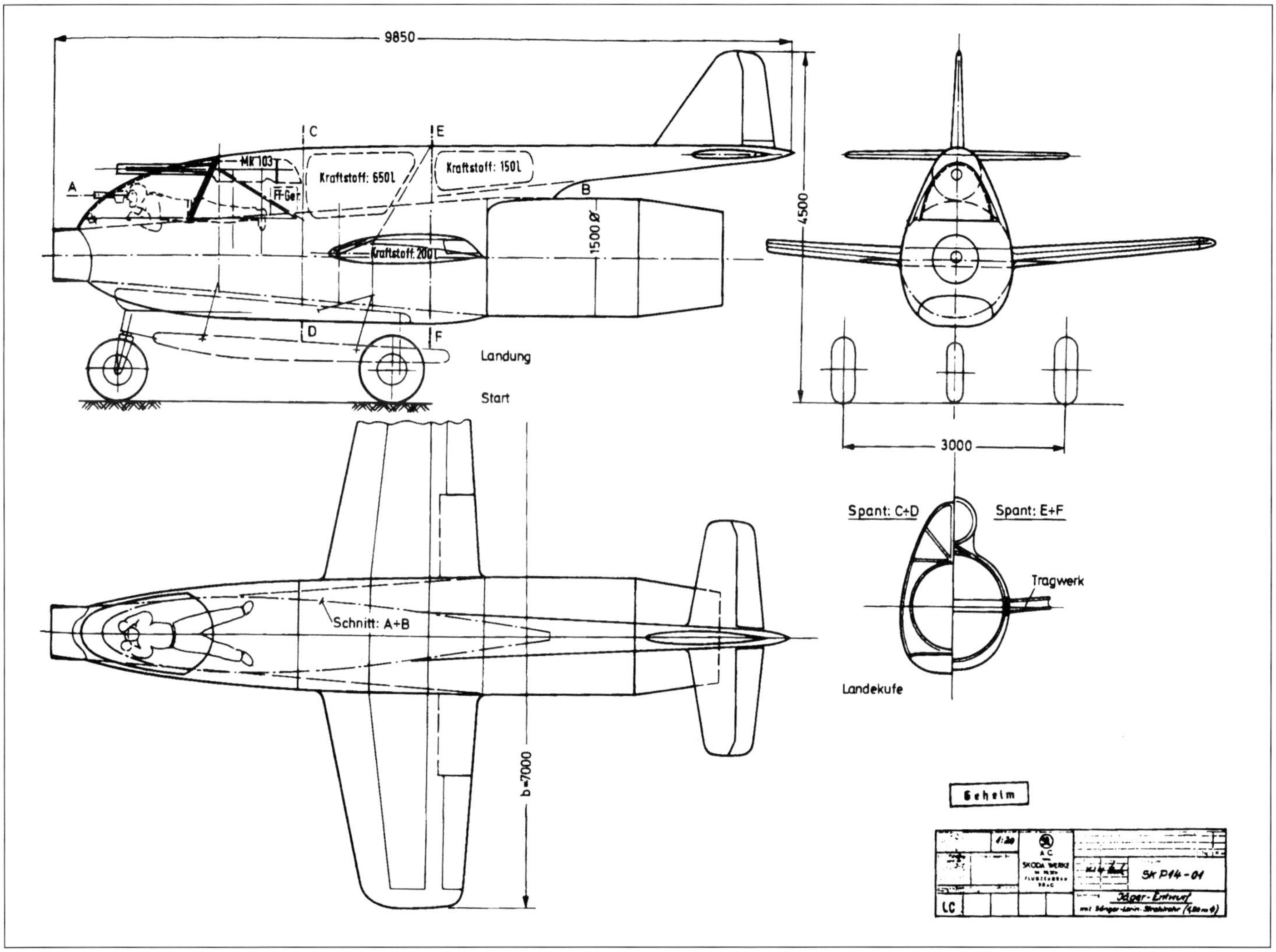

Design drawing for the Skoda-Kauba Sk P 14, an improved variant of the ramjet fighter from 25 February 1945.

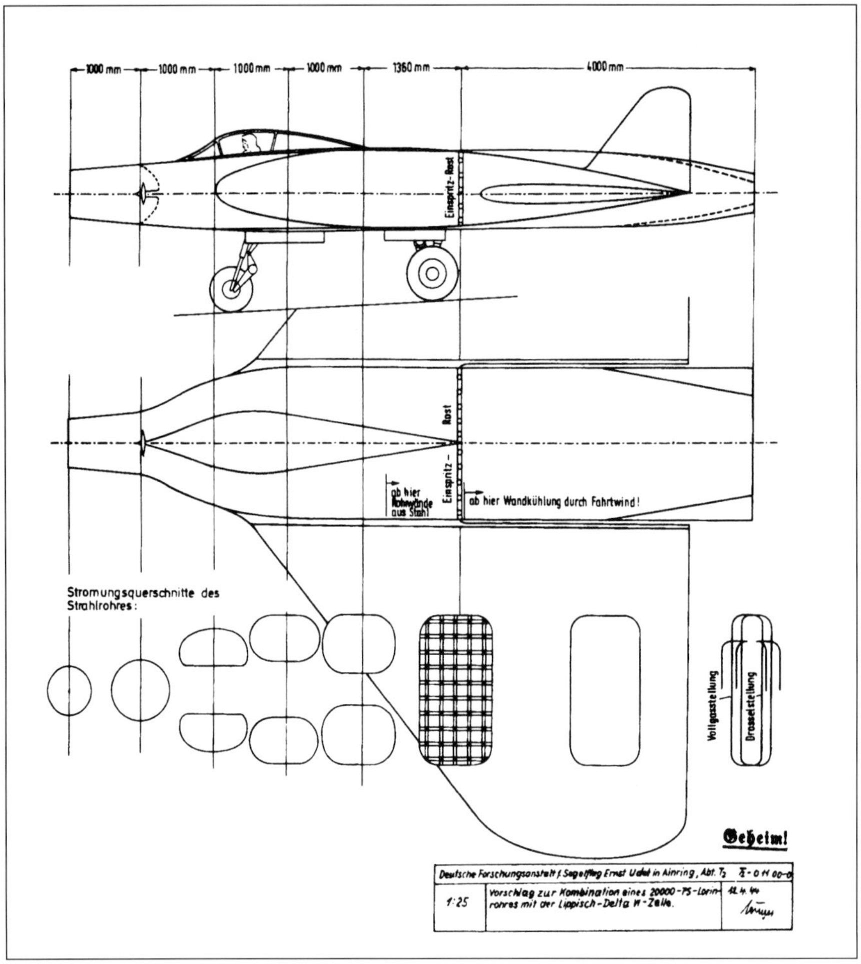

Proposal sketch by Eugen Sänger for installing a 20,000 hp ramjet engine in the Lippisch "Delta-Projekt", from 12 April 1944.

Yet another incomparable model by Günter Sengfelder. The Focke-Wulf "*Triebflügel*" (Powered Wing) shows that, while carrying the same name as the Lippisch design, this was an entirely new concept.

To add to this, it must be mentioned that Professor Eugen Sänger had drawn up initial ideas for an atomic rocket as early as 03 June 1944, a theme that he pursued incessantly the rest of his life.

Eugen Sänger died much too early on 10 February 1964 from a heart attack at the age of 59. How little understanding was given to his works at the time is evidenced by a hastily handwritten comment scrawled at the edge of one of his test reports: "These idiots, at these speeds you can't even fly or aim." This comment says more than any words and once again shows the ignorance and inability to fully grasp and make use of the unique technological understanding at the time.

Building upon the experiments with ramjet propulsion, Sänger, in cooperation with the Skoda-Kauba Works, drew up a concept study for a ramjet fighter. Two variants were sketched, both similar in layout with the fuselage effectively comprising the ramjet with wings and empennage attached (see picture on page 184). As a rule, to power a ramjet carbon hydrogen fuels such as diesel or gasoline were used, giving a combustion temperature of about 1,800% C. However, in order to increase efficiency, fuels based on alloy dispersion such as aluminum or magnesium were also considered with an eye toward increasing combustion temperatures up to 3,000% C. In the night fighter role, special fuels like alcohol would be used to keep the exhaust flame invisible. However, since the headaches involved in procuring adequate quantities of fuel during the second-to-last year of the war were painfully obvious, and prospects for any future supplies were dim, eyes were kept peeled for viable alternatives. As we all know, necessity is the mother of invention, and it was Professor Eugen Sänger who came up with the idea of using coal granules in special holding grates as fuel. In conjunction with this, successful experiments were conducted between 10 October and 24 November 1944 using various types of coal. The method was improved by spraying the coal granules with heavy oil, which led to better combustion results and ensured a continuous thrust output. Between 21 and 23 February 1945 negotiations were carried out with the Skoda Works with an eye toward using this method for the planned ramjet fighter. These plans, however, fell by the wayside, for by this time the Skoda Works were no longer capable of carrying out such an expansive undertaking.

Parallel to the efforts of Professor Sänger, a Dr. Schwabl in Vienna was also wrestling with the use of coal granules in several promising experiments with the DFS as a replacement fuel. After Professor Sänger proposed equipping the Lippisch project with a ramjet as well (see drawing left), Alexander Lippisch instigated ideas of fitting his supersonic delta with a ramjet that would occupy virtually the entire wing area; thus he, too, named his product a "*Triebflügel*" (Powered wing). However, his *Triebflügel* had nothing to do with the other, better known *Triebflügel* project by Focke-Wulf (see photo left).

After an exchange of ideas with both Sänger as well as Schwabl, it was decided that Lippisch's ramjet would be powered by refined coal and injected heavy oil. It was expected that a load of 500 kg of granulated coal particles would offer a flight time of at least 30 minutes. By means of pressurized air, the coal granules would be measured out similar to a pneumatic feed system via a rotary airlock valve, blown into the combustion chamber and there mix with the heavy oil also being sprayed into the chamber. Since Professor Sänger's experiments in this area were by far the most advanced, his experience with this fuel mixture would influence the design conception of the Li P 13. For takeoff and for reaching the speed needed for operation of the ramjet, an additional rocket motor would be installed in the aft section of the Li P 13. Planned armament included the standard MK 108 machine cannon common to high speed fighters, with a caliber of 20 mm. Takeoff would be by means of a takeoff cart that would remain on the ground once the airplane was airborne; a landing skid would be deployed for landing.

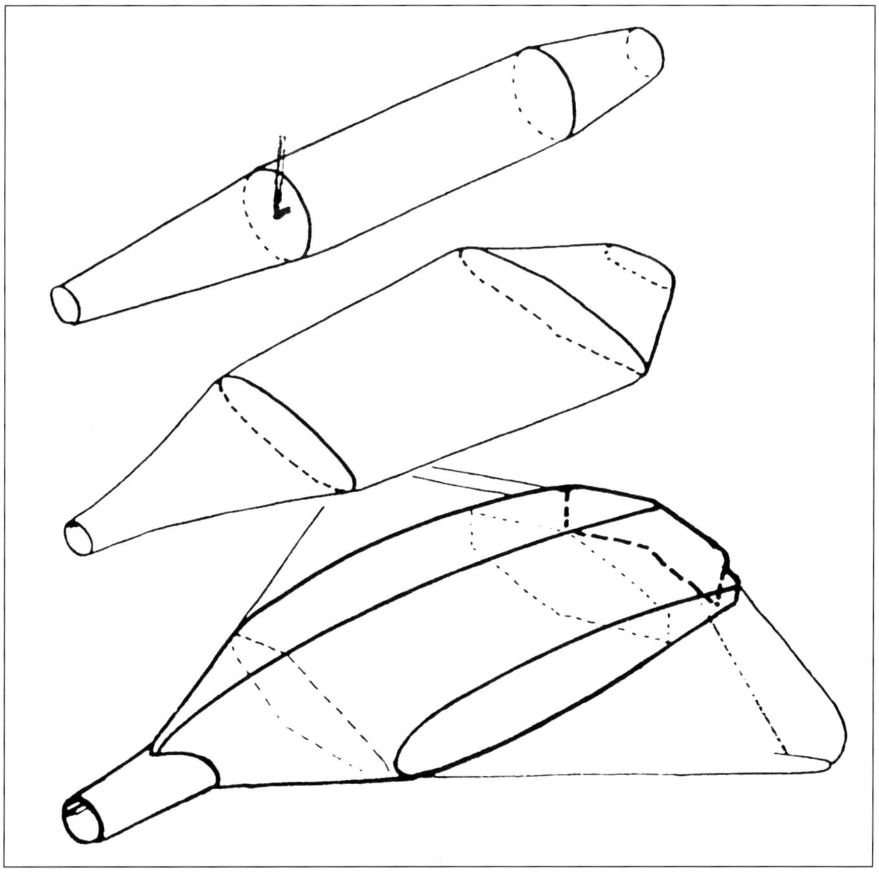

This drawing shows the developmental stages from ramjet to "*Triebflügel*".

View of the DM 1 from the front shows once again the results of a triangular wing design and the triangular rudder.

This detail photo of the nose section once more shows that the lower nose has been fitted with a section of plexiglass to improve pilot visibility.

The DM1 at the airfield of Prien am Chiemsee, shortly after the airfield was occupied by American troops.

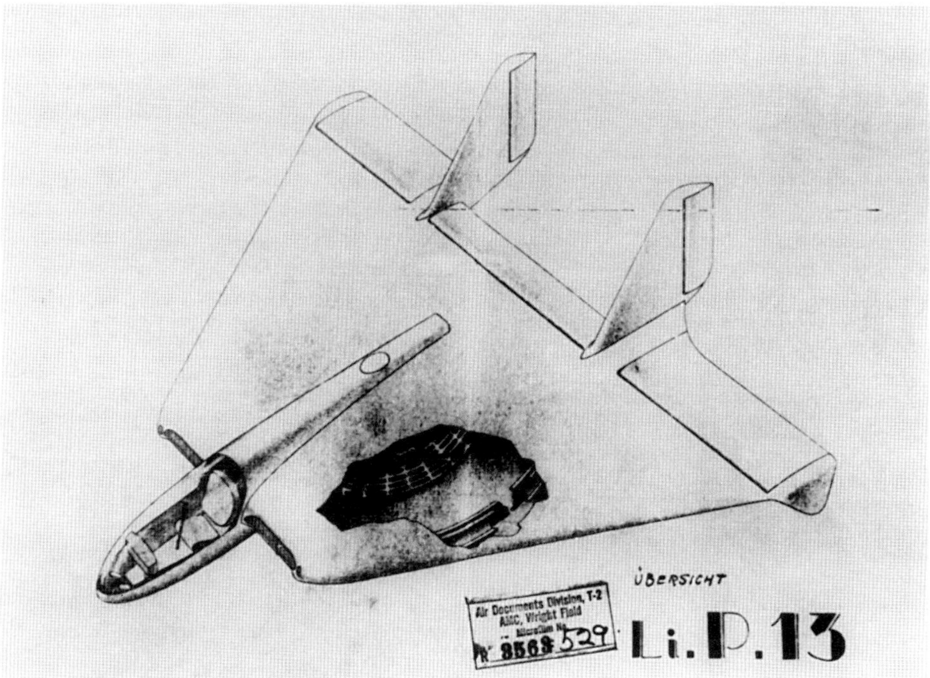

Overview design drawing of the Lippisch Li P 13, a planned variant of the Li P 13 with a much larger engine area and a kind of rotary airlock valve to feed the granulated coal particles into the combustion chamber.

The DM 1 seen once again from the side shows what a jump in development the DM 1 was when compared to the American DC-3 in the background.

Like the Brothers Horton, Lippisch envisioned an extreme sweepback of the wings for his project, with an angle of 60%, which he also used for the rectangular rudder. Lippisch put the pilot seated inside the vertical stabilizer, thereby providing him excellent visibility. With a wingspan of just 6 meters and a length of 6.70 meters, the relatively small craft would have been quite maneuverable in combat; combined with the pilot's position giving him a view of 'everything in sight', these would have been advantages of inestimable value for a successful dogfight.

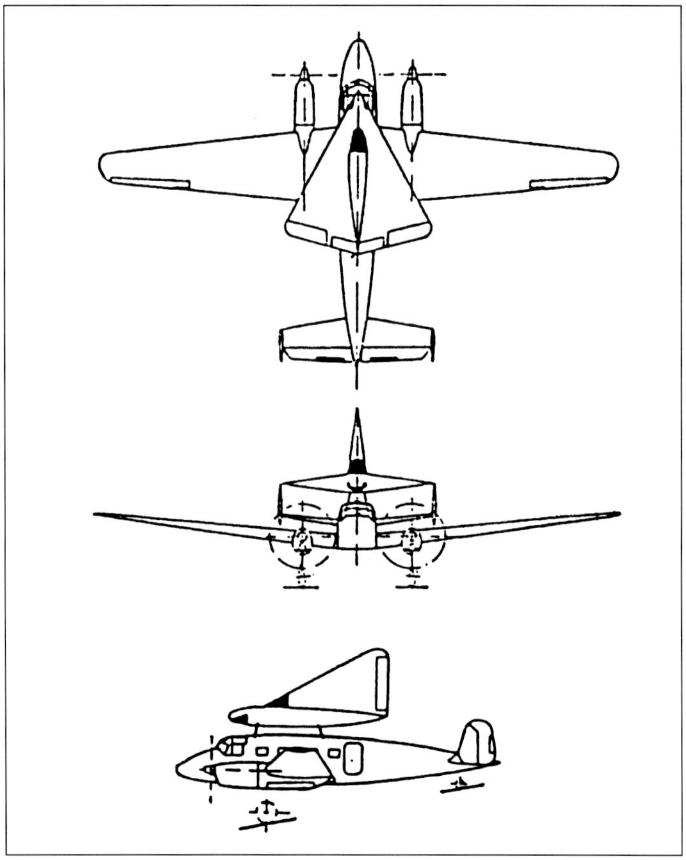

Three-view sketch of the piggy-back combination of a Siebel Si 204 and the DM 1.

Lippisch's concept combined with Sänger's coal-powered ramjet offered excellent prospects for success and was tackled with great expectations and hope. As early as May 1944, when the question of propulsion was still open, successful tests with free-flying models were carried out on the Spitzenberg mountain near Vienna. Alongside these tests, the *Aerodynamische Versuchsanstalt* (AVA) conducted wind tunnel experiments at Göppingen, which also confirmed the viability of the system. In consequence, it was decided to proceed with the construction of an unpowered prototype made of wood in order to test the actual flight handling characteristics. This test prototype, which actually should have carried the designation of Li P 13 V1, was called the DM 1. This designation was chosen to convey the idea that the project was only a prototype of the academic flying groups, or AKAFLIEG, Darmstadt and Munich (hence DM), and also served simultaneously to maintain secrecy.

In order to understand this in context, we must delve deeply into the history of the DFS. This does not mean that an entire history of the beginnings of the DFS (*Deutsche Forschungsanstalt für Segelflug, e. V.* - German Research Institute for Unpowered Flight) will be examined, but this much needs to be said: during its early days the organization was stationed in Darmstadt, and Professor Alexander Lippisch was one of its founding fathers. After he left the DFS, Lippisch did not lose contact with his old comrades-in-arms; he was also one of the ones to foster development of the academic flying groups and to keep them operating. This was undoubtedly one of the reasons that he arranged for the construction and flight testing contracts for the DM 1 to be awarded to the *Flugtechnische Fachgruppe* (FFG, or Flight Technical Studies

Group) Darmstadt, assisted by the activities of the Darmstadt 'academic flyer' Wolfgang Heinemann. In order to carry out this massive undertaking as quickly as possible, the team linked up with the AKAFLIEG Munich after being bombed out on 11/12 September 1944, resettled in Prien on the Chiemsee and there began to work in the FFG Munich's hangar facilities. The spiritual patron of the DM 1 - Alexander Lippisch - on the other hand, was no longer involved in the DM 1's further course of construction.

For flight trials the DM 1 was to be carried to altitude piggy-backed to a twin-engine Siebel Si 204 A and then released. Two factory-fresh machines were sent to Prien, where they waited patiently for their missions. The DM 1 was fitted with a simple fixed tricycle undercarriage for landing. In order to reach the hoped for speeds in flight, it was planned to attach solid-fuel rockets to the mounting points for the piggy-back configuration on the main undercarriage legs.

Hans Zacher, who also served as the testing director for the Ba 249 *Natter* manned anti-aircraft rocket, was to have taken over flight testing for the DM 1. He often came to Prien during the DM 1's construction phase and carried out several cockpit tests on the ground. On 3 May 1945 American troops occupied the airfield at Prien, bringing all dreams of testing to an abrupt end. Nevertheless, the Americans decided to let construction of the DM 1 continue to completion under their supervision. Once finished, the DM 1 was packed into a transport crate and sent to the US. There it was subjected to intensive wind tunnel testing, became the pattern for a wide variety of postwar delta wing designs and now waits for better days - quite the worse for wear - in the warehouse of the Smithsonian National Air and Space Museum in Washington (here, too, there's talk of restoration in Germany sometime in the near future). Thus, before it really ever began, also ended the story of the first German supersonic fighter in a fully unspectacular fashion.

For those who would like to build a realistic model of the Li P 13, the 1/48th scale kit from Revell comes highly recommended, with virtually no room for complaints. Unfortunately, we must now bid adieu to this most interesting project in order to turn our attention to yet another novelty in the history of aviation - the first variable-sweep wing design.

Group photo with the crew. The DM 1 with a few undaunted supporters and co-workers on the DM 1 project: (l to r) Herbert Dieks, Wolfgang Heinemann, Klaus Metzner and Hermann Nenninger.

Three-view comparison of the Lippisch Li P 13 a and the DM 1.

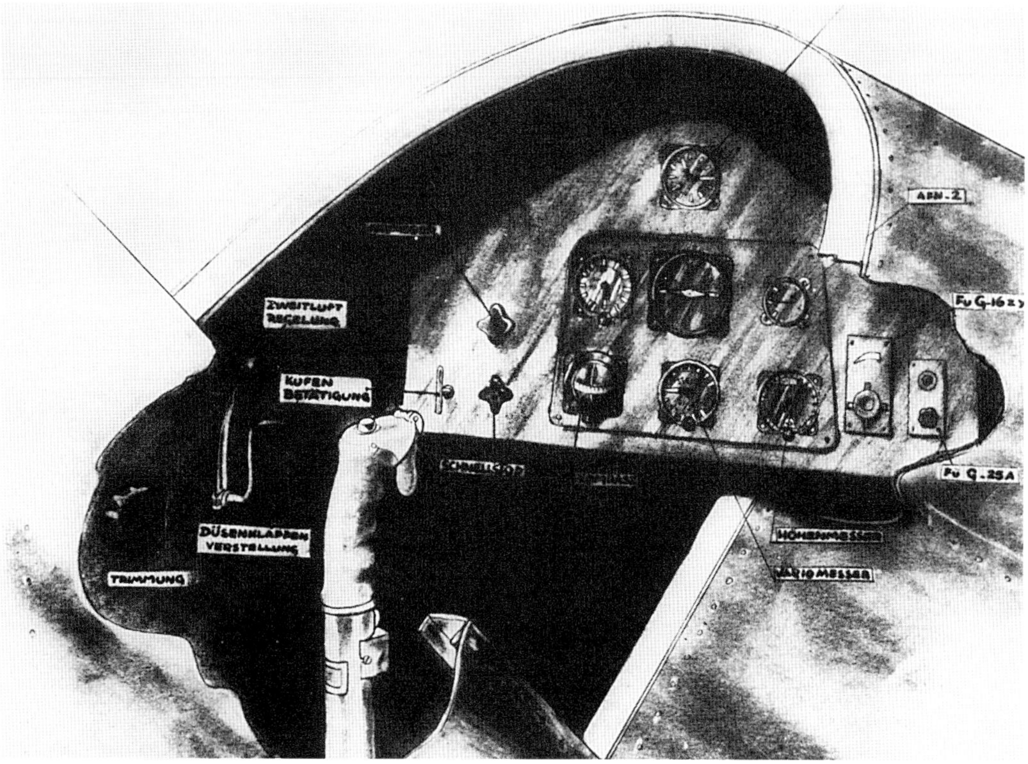

Kept to the absolute minimum: the instrumentation in the cockpit of the Li P 13 a.

As an aside, it should be noted that the Americans are presently carrying out hypersonic experiments with the X-43 test bed which, similar to the Li P 13 project, makes use of an engine combination consisting of an accelerator rocket motor to boost the plane to the operating speed for the main ramjet engine, a scramjet (supersonic combustion ramjet). The X-43 is carried beneath the wing of a B-52 to altitude, released, and with the assistance of three Pegasus rockets, accelerates to supersonic speed. Once the Pegasus rockets have burned out, the ramjet pushes the aircraft to speeds of up to 10,000 km/h. As once envisioned by Professor Sänger, a ramjet is being used to accelerate a flying body to speeds in the hypersonic region. Not until now, some fifty years later, have we advanced to the point technologically where we can turn Professor Sänger's ideas into reality. It remains to be seen when the ion propulsion system and possibly the atomic rocket will one day be so far advanced that we can find a practical use for these as well. Professor Sänger had the misfortune of being so forward thinking with his ideas at the time that the technological means of the day could not keep up and, after the war, he found those ideas cast by the wayside. There was no one to lobby for them.

Messerschmitt Me P1101 V1 (Swing-wing)

Professor Willy Messerschmitt wanted to make his own mark on the 'sound barrier' phenomenon and, in so doing, just happened to pave the way for the development of today's swing-wing aircraft. A little known bomber project with 'swing-wings' was drawn up in April 1944 under the project number Me P 1101. The four-engine bomber was to have been powered by Heinkel He S 011 turbine engines and have a bomb payload capacity of three metric tons. The wings could pivot from 8% to 47% sweep, giving it a span varying from 14 meters to 20.50 meters. Project blueprints for the P 1101 - XVIII 103 were completed by 29 June 1944 that give a good overview of the state of the project. Judging from the precision work of the drawings, it appears that a considerable amount of detailed preparation went into the effort beforehand. From a fairly early stage it was apparently obvious that this advanced bomber design had virtually no chance of coming to fruition, and because of this the project was dropped in favor of a similarly de-

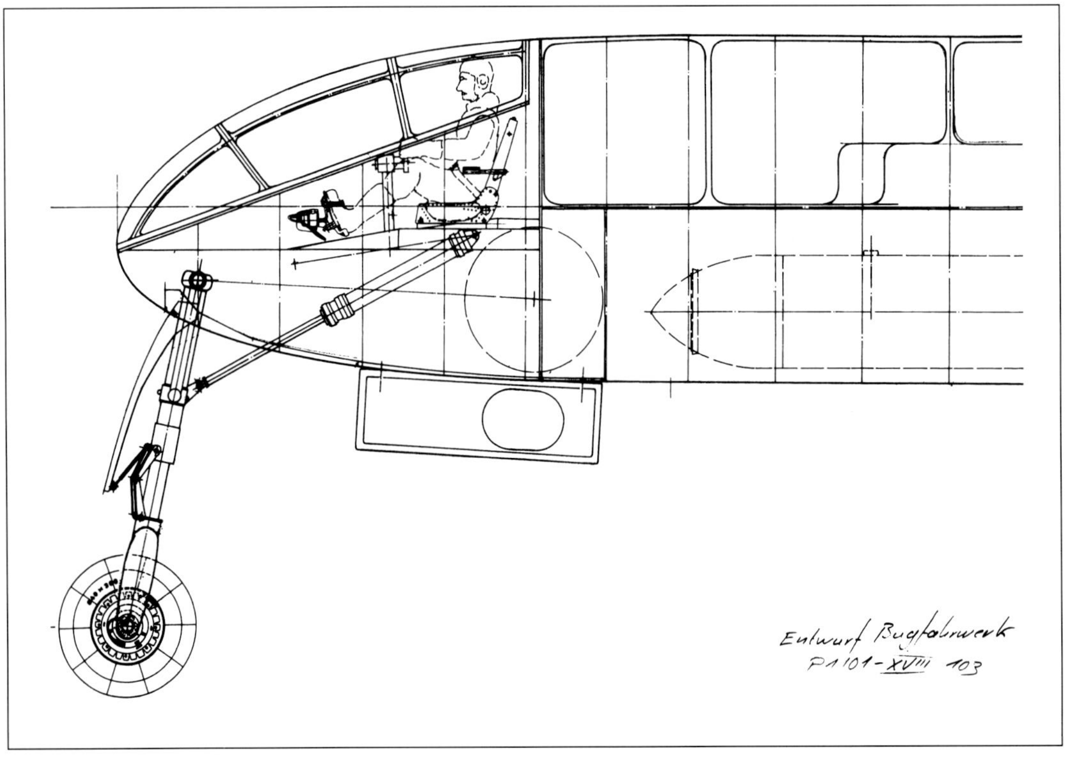

Detail drawing of the nose gear on the Me P 1101-XVIII 103 based on the original, copied by aircraft undercarriage specialist Günter Sengfelder.

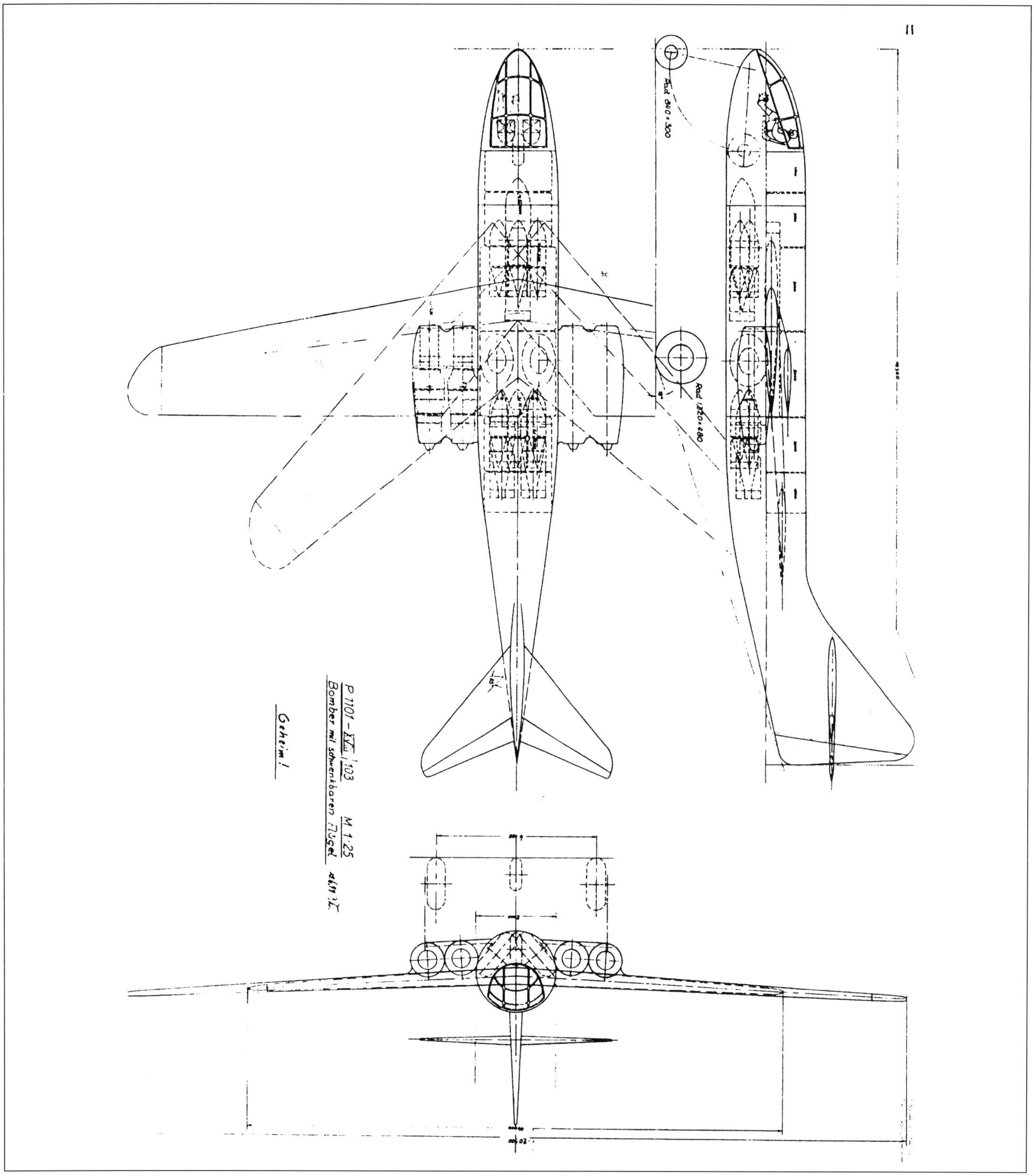

Project drawing for the Messerschmitt Me P 1101-XVIII 103, from 29 June 1944. With its advanced lines, the design could pass for one of today's modern aircraft.

The Me P 1101 V1 at Oberammergau, shortly after the Americans took over the site. An American G.I. sits on the wing. An unshrouded Junkers Jumo 109-004 B jet engine can be seen to good effect beneath the wing.

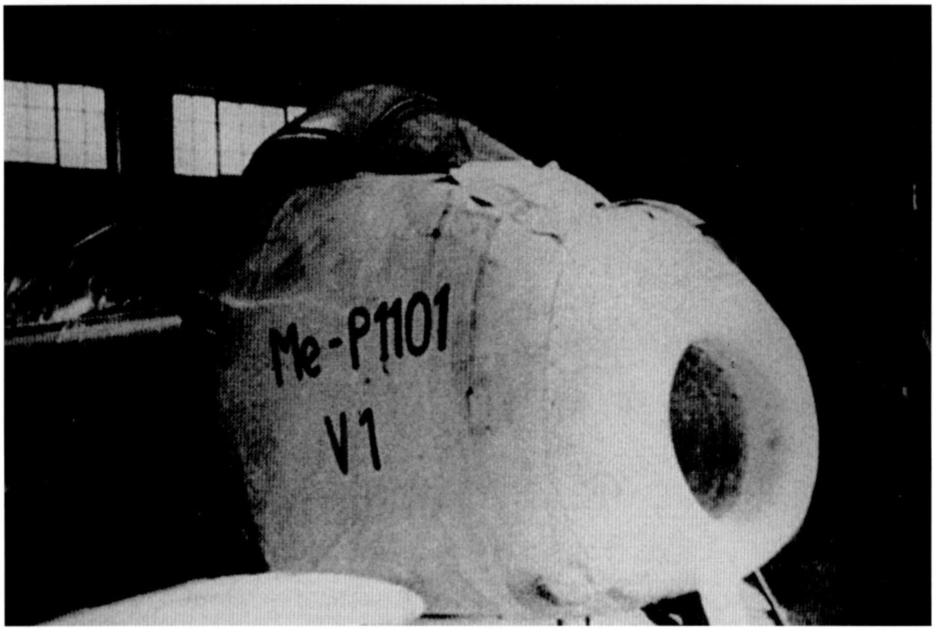

This photo shows the repair work being carried out on the damaged front fuselage. The replacement plates for the fuselage skin are quite visible where they've been applied.

The Messerschmitt Me P 1101 V1 somewhat worse for wear at the Messerschmitt hangar in the Oberammergau Works. The obvious damage to the forward nose section was most probably caused by mistreatment at the hands of American ground troops when they pulled the Me P 1101 V1 from a cave where it had been hidden.

Technical Data:

	Experimental aircraft	Production aircraft
Wingspan	8080 mm	8250 mm
Wing area	15.80 m^2	15.85 m^2
Length	8835 mm	9175 mm
Max. height	approx. 3930 mm	4064 mm
Nosewheel	465 x 165 mm	500 x 180 mm
Main wheels	660 x 190 mm	740 x 210 mm
Engine	Jumo 004 B	He S 011
Sweep angle	35-45%	40%
Takeoff weight	2841 kg	4064 kg
Takeoff run		750 m
Climb rate at 0 km		22.2 m
Max. speed at 7 km		985 km/h
Max. speed at 6 km	1100 km/h	
Max. speed at 0 km	1050 km/h	
Landing speed		172 km/h
Rollout distance		570 m
Armament	(2 MK 108)	4 MK 108
Ammunition	(200 rounds)	400 rounds
Status	11/10/1944	2/22/1945

Technical datasheet for the three-view on the following page.

signed fighter project that retained the P 1101 project number. It's a shame, really, that this elegant design never got the attention it deserved. It wasn't until the American B-1 that a swing-wing project of similar dimensions was realized.

Following the saying that "only fools rush in where wise men fear to tread", Professor Messerschmitt opted for an experimental aircraft that would first test the optimum wing sweep in actual flight. To this end, the wings would be preset on the ground at three different sweep angles: 35%, 40%, and 45%.

In July 1944 he proposed to the RLM the construction of such an experimental design, which would provide basic information for a full scale production program based on the concept. By 10 November 1944 a complete project description with performance characteristics was published under the project designation of P 1101. Like the testing of the Ju 287, efforts were made to incorporate as many parts from existing designs as possible with the goal of building and testing the experimental aircraft with as little time expenditure as possible. Primary consideration was given to using components from the Me 262 and Me 109, such as the complete wing section from the production Me 262 with only minor changes. Furthermore, the idea was to build several experimental aircraft in order to ensure testing would continue uninterrupted. Initially, plans called for the use of a Junkers Jumo 109-004 B turbojet engine, but subsequent conversion to the more powerful He S 109-011 Heinkel turbine would be possible without difficulty.

Once the project was established as such, over the second half of December 1944 work began on starting construction of individual components for the first prototype of the Me P 1101 V1.

By the end of January 1945 the production of the aircraft was in full swing. Three wing sets, with sweeps of 35%, 40%, and 45%, were built to facilitate later flight testing. The aircraft would first be tested with a sweepback of 35%, with subsequent experimental flights making use of the 45% sweepback. Only in the event that these two wing sweeps failed to provide the anticipated results would test flights be carried out using the wing set with the 40% sweepback.

Parallel to construction of the experimental aircraft, under the direction of Waldemar Voigt and Hans Hornung the project group worked on drawing up the blueprints for the later production version - time was of the essence, whatever the cost. It must be kept in mind that all this occurred at a time when there was not as yet any definitive results from flight testing - the test plane hadn't even been built yet. In addition, they had no idea what the contractor (i.e. the RLM) actually wanted. This rather amazing set of facts can only really be understood in light of the increasingly hopeless conditions Germany faced as the war drew to a close.

Despite the hostile conditions, the workers at Messerschmitt succeeded in completing the Me 1101 V1 at the Oberammergau Works before the Americans occupied the grounds on 29 April 1945. Assembly of the second test prototype, the Me P 1101 V2, was well advanced at this stage as well, so that the Americans also discovered many of this aircraft's completed subassemblies.

Shortly before the end of the war the Me P 1101 V1 was moved to a nearby cave. It was there that the American infantrymen found the experimental plane and brought it into the light. Either because of the improper handling by the soldiers or perhaps due to a last minute effort by Messerschmitt personnel to destroy the airplane, the Me P 1101 V1 suffered some minor damage, primarily in the nose section (see photos left).

At first the machine was set up in the open, photographed, and then moved to Hangar 615 at the Messerschmitt Works in Oberammergau. There, on 7 May 1945, American specialists from the CAFT (Combined Advance Field Team), first came upon the aircraft. On 17 May 1945 additional experts from the USSTAF-ATI (United States Strategic Air Forces-Air Technical Intelligence) arrived on the scene, and a final team of experts - this time from the Bell Company - showed up on 21 May 1945 under the supervision of Robert J. Woods.

It wasn't long before the project was included in the category of worth continued development, and work immediately began on repairing the damage on site (see photo page 196). Once the machine had been restored enough to be transported, it was crated up and, along with any documents the personnel could get their hands on, shipped to Wright AFB (Air Force Base) in the US, the Mecca of the later War Prize display at Wright Field. Any plans for bring-

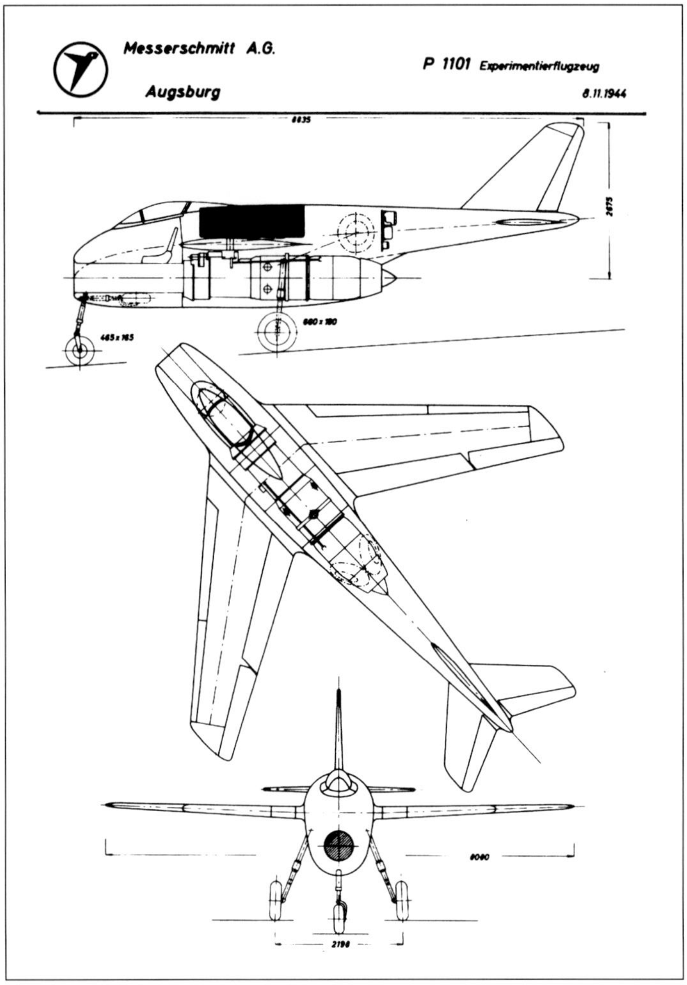

Three-view of the Me P 1101 V1 as was planned as an experimental design by the Messerschmitt Works on 11/8/1944.

ing the Me P 1101 V1 to a flyable state in Germany were dropped when several unexpected problems cropped up (some of the plans had made their way to Paris and the French were adamantly opposed to giving these to the Americans).

In 1949 the Bell Company once again tackled the Me P 1101 V1 project, tweaked the design so that the wings could now be repositioned during flight, and called the whole thing the Bell X-5. Under this designation the first prototype took to the air on its successful maiden flight on 20 June 1951 with test pilot Jean Ziegler at the controls. The aircraft subsequently began a rigorous flight test program. On the fifth test flight the wings were swept in flight for the first time. The second prototype also entered flight testing, on 10 December 1951. Although the second prototype crashed during testing, killing pilot Major Raymond Popson, the testing program continued through to the end of 1955. At the conclusion of the test program the surviving Bell X-5 was taken to the US Air Force Museum at Dayton, Ohio. During the Bell X-5's testing program, original parts of the Me P 1101 V1 were used for static experimentation. But once the testing had been completed, these original components suffered an ignominious fate, going the way of all mortals - they wound up on the scrap heap.

With this sobering thought, there remains but one additional summarization in the matter of wing sweepback. What is left of German wing sweep development that had begun with such high hopes and was carried along with great enthusiasm? Before we conclude with this subject, awaiting us is an exciting 'second helping' - the story of the Junkers Ju 322.

The excellent model by Mr. Günter Sengfelder in 1/15th scale shows the Me P 1101 V1 being prepared for a fictional test flight.

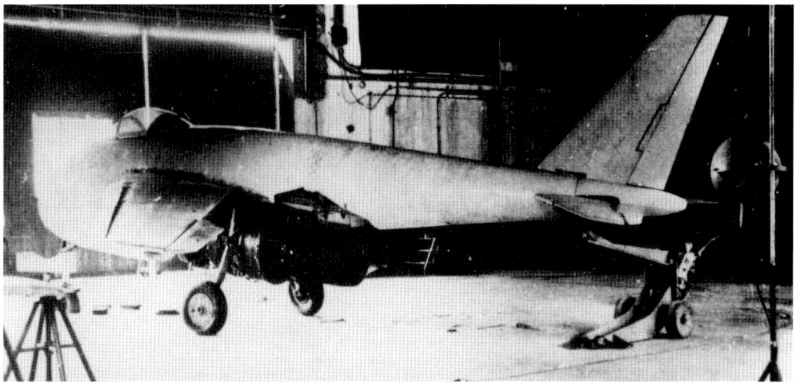

The following series of pictures shows the experimental airplane from several different angles in the Messerschmitt Works Hangar 615 at Oberammergau. This view from an oblique port angle offers a good view of the engine installation beneath the fuselage.

This view from behind shows to good effect the modern design of the Me P 1101 V1.

Side view of the Me P 1101 V1. The low-profile, low-drag canopy was tested on the Me 262 V9 as a so-called "racing canopy" and reputedly enabled the Me 262 to attain higher speeds.

Posing in the open in front of the hangar at Oberammergau, this shows a front view of the Me P 1101 V1.

The Me P 1101 V1 as a 1/15th scale model by Günter Sengfelder on the "runway".

Like peas in a pod: the Bell X-5 with registration number 01838 was the first test version of the Me P 1101 V1 copy to be flown in the US. It can still be seen in a museum today. (Photo: USAF)

Air-to-air photo of the Bell X-5. This is probably what an aerial shot of the Me P 1101 V1 would have looked like had it ever flown. (Photo: USAF)

This photo shows the Bell X5 with the three optional wing sweep positions. These are the same positions for the Me P 1101 V1, although the wings had to be set on the ground. (Photo: USAF)

In conclusion, yet another shot of the Me P 1101 V1 as a testbed with an American Allison J-35 jet engine fitted.

The Non-swept Wing Junkers Ju 322 and its Unrecognized Qualities

Whoever looks at the modern transports such as the Lockheed C-5A Galaxy or the Antonov An 225 will quickly note that both have swept wings. Now, you can be forgiven for saying that this is nothing dramatic and probably is necessary, since after all both airplanes not only carry heavy cargo loads but also fly at near sonic speeds. However, the swept wing does not justify its existence solely because of its high speed potential; there is another reason behind its use, as is demonstrated by the Junkers Ju 322 - a design considered to be a failure in the annals of aviation history.

In order to provide a detailed account of the events surrounding the design, we must first explore history in more depth. The Second World War was in its second year. Most surprisingly, the war against France concluded in short order with the Third Reich as its victor, and there was much speculation on how things should proceed from there. There wasn't even the slightest preparation for a war against England involving the necessary landing operations. This resulted in the hastily prepared Operation *Seelöwe* (Sea Lion), the naval landing on the British Isles. The operations plan called

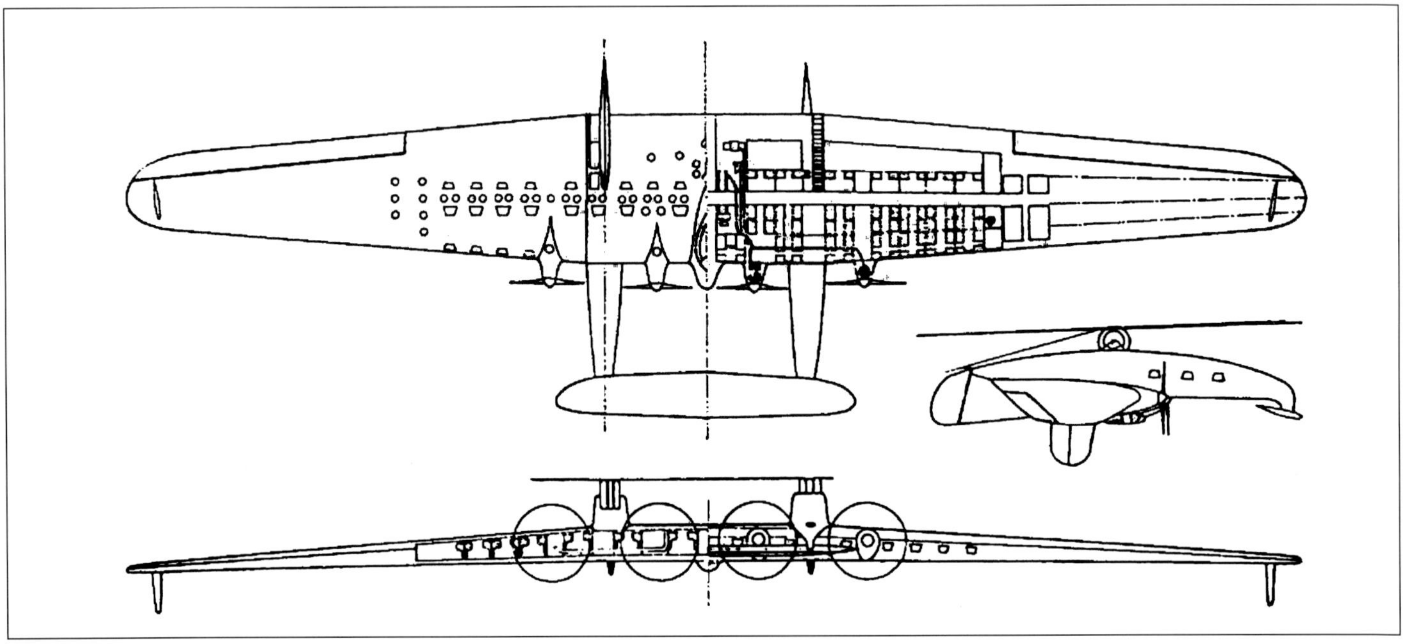

This three-view of the Junkers J 1000 from 1924 shows the unusual planform of the passenger design, with a giant 80 m wingspan. This was the first design to carry both passengers and cargo in the wings.

for a landing at six different sites along the English coast with the intent of establishing bridgeheads at those points. An absolutely insane plan, given the fact that a fully intact, highly superior British Home Fleet would have to be dealt with, a fleet with which the weak German sea power had nothing to compare. The prerequisite for this crazy operation was the absolute air supremacy over the entire operations theater. But the so-called Battle of Britain very quickly revealed the limits of the German *Luftwaffe*, and on 12 October 1940 Operation *Seelöwe* was quietly shelved. But you would be mistaken to think that this signaled the end to any plans for a landing on the British Isles. If it wasn't possible by sea, then it would have to be by air - despite the bad experience during the Battle of Britain. In spite of everything, it was felt that there was enough strength to conduct an airborne landing in a restricted area and guarantee air superiority in this area for a limited period of time. From these considerations was born the requirement for rapid development of a superlarge transport glider that would be able to carry a large quantity of men and material into the operations area in a short amount of time. This superheavy transport glider was to have been designed to be disposable - for one-way operations only. The entire operation was given the designation *Warschau* (Warsaw). Both Messerschmitt and Junkers each were given a contract for creating such an aircraft, with Messerschmitt primarily making use of steel in its design. The Junkers Company, on the other hand, was given the task of using chiefly wood in its design as an alternative in the event that the war would restrict the supply of steel. So much for its history.

In the following description the Me 321 *Gigant* (Giant), Messerschmitt's design, is not discussed at all, primarily because there exists plenty of literature, specifically a variety of detailed publications. Load requirements for these heavy lift gliders ranged from the ability to carry a Panzer IV, an 88 mm anti-aircraft gun with its tow vehicle, plus up to 120 - 200 fully equipped soldiers. As the basis for such a flying machine the Junkers Company dusted off the altogether positive experience it had enjoyed with the Junkers Ju G 38, the largest land plane of its day.

As early as 1924 Hugo Junkers had attracted considerable attention with his Junkers J 1000 project, an enormous machine with an 80 m wingspan and a canard layout, i.e. with the empennage/control surfaces protruding ahead of the wings when viewed from the direction of flight (see drawing on left). It was capable of carrying up to 100 passengers in luxurious cabins, all inside the spacious wings. This idea by Hugo Junkers, of a 'thick, load bearing wing', was used as the basis for the Junkers G 38 mentioned earlier. This principle proved to be quite successful on the Junkers G 38, so that the first project sketch for the heavy lift glider, the Junkers EF 94, also envisioned a gigantic wing with thicker center section that would serve as the cargo compartment - similar to the passenger compartment in the wings of the Junkers G 38.

Professor Heinrich Hertel, who at the time was the technical director at Junkers, and his colleagues, *Dipl.-Ing.* Gropler, director of development, and *Ing.* Paul J. Hall, responsible for detail design, used the EF 94 as the basis for developing the actual Ju 322 glider, designated by the RLM as the 8-322.

The Ju 322 had the legendary wingspan of an unbelievable 82.35 m (other sources report it as 61.70, 62.35 m, or give contradictory data on the span), a high aspect ratio wing and a thick, lift-generating profile - something which would later have drastic con-

This drawing of the Junkers J 1000 at an airfield demonstrates yet again the imposing size of the planned passenger liner.

The Junkers G 38, at the time the largest land-based plane in the world, was the first successful step toward the "cargo-carrying wing" project, something that was later realized with the Ju 322.

sequences. Compared to the wing, the fuselage appeared narrow and compact, serving only as a support for the empennage. With some limitations, the design could be said to be a "flying wing", something of a paradoxical designation since it was fitted with a completely standard empennage.

Basically, the Ju 322's layout was nothing more than the aforementioned giant wing, whose center section served as a cargo hold for a 20 metric ton payload. Loading and unloading was accomplished by removing the wing leading edge. The cockpit was offset to port, alongside the cargo bay. The first prototype, the Ju 322 V1, was also fitted with two cylindrical ballast containers, one each to the left and right of the cargo compartment, which could each hold about 2,000 liters of water. These had erroneously been reported in earlier publications as gun positions, fitted with 7.9 mm MG 15s. This error is undoubtedly due to the fact that Gert W. Heumann had modified the only known original photo to date in such a way that

This photo clearly shows that at least some of the passengers were carried in the thick wings of the Ju G 38. The performance of the Ju G 38 also made quite an impression on the Japanese, and they built the type under license as the Mitsubishi Ki 20 super-heavy bomber (German bomber variants of the Ju G 38 carried the designation Junkers K 51).

two gun positions can clearly be seen (see picture, page 211). It is purely speculative whether or not such armament was ever actually envisioned for the Ju 322. Apparently, these guns would have been used on approach to the operations area to keep the heads of enemy ground troops down as well as to provide covering fire during cargo/ troop "debarkation". Later plans called for up to five such gun positions. All control surfaces were enhanced by power boosters as it would have been impossible to manipulate the control surfaces exclusively by muscle power alone, given the enormously high control pressure.

Loading took place via an inclined ramp, since at that point the Ju 322 would already be resting on its takeoff trolley. However, provided it was designed to be stable, this trolley would be capable of supporting a fully loaded Ju 322, weighing in at some 45 metric tons. Variants were designed with 8, 16 and even 32 wheels, which led to the weight of the takeoff trolley climbing to 8 metric tons, believe it or not. It was planned that, once the Ju 322 reached takeoff speed, the trolley would separate from the glider and roll along the ground until coming to rest. In order to test the takeoff method several trials were carried out using a model, which revealed that

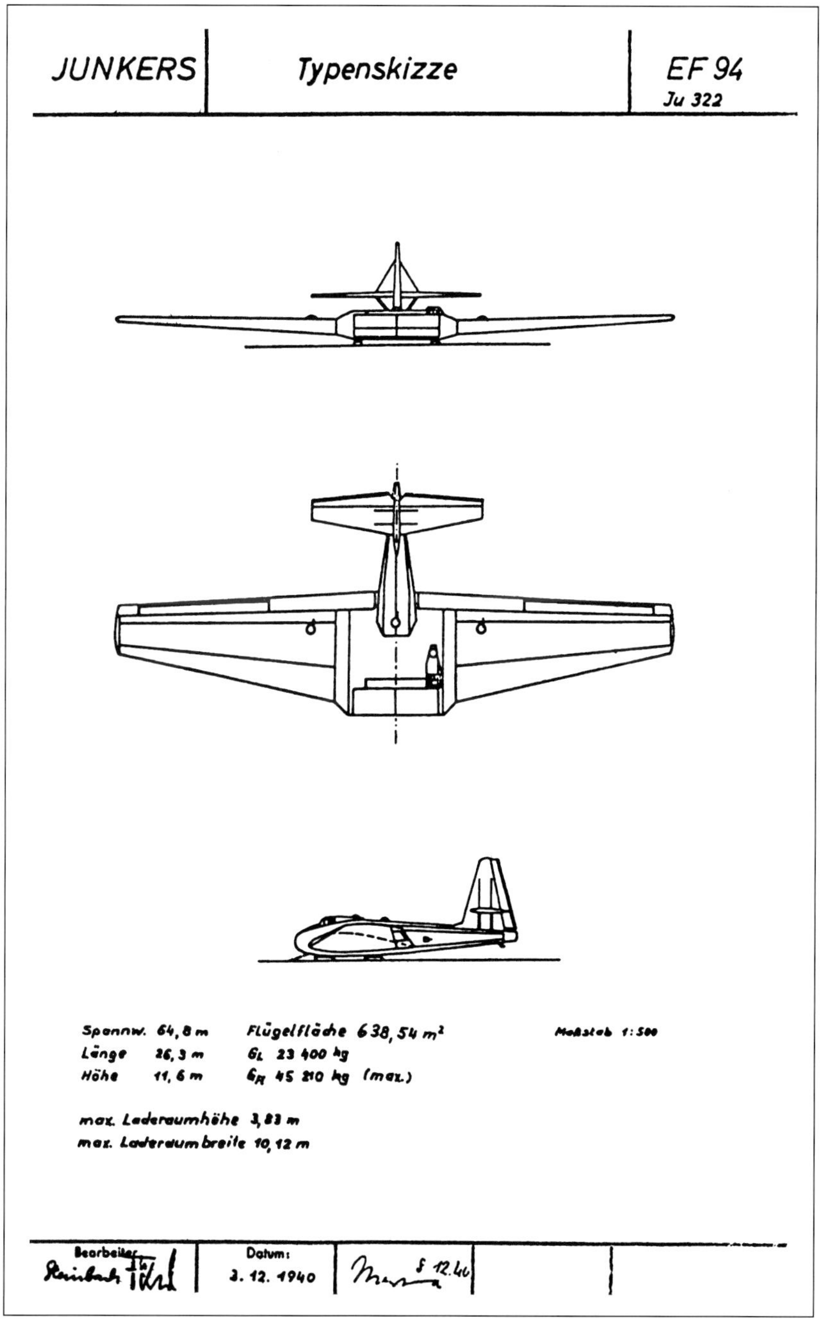

The commonly found three-view of the Junkers Ju EF 94 of the project study on the Ju 322 shows a massive wing with short fuselage and extra large rudder.

this method was ultimately perhaps not the wisest approach. Since, however, there was currently no better takeoff method available for this colossus, it was simply tested as was. With limited success on its first attempt, as we shall see shortly. The Ju 322 would land on four massive skids. To shorten the landing run, a brake shaft was developed that could be extended electrically; the brake shaft also served as an aid to loading by keeping the tail of the empty, tail-heavy Ju 322 raised.

Since the cargo glider was to have been built entirely of wood, an enormous amount of wood had to be acquired for the 100 aircraft planned for construction (originally there was even talk of building 200 machines). It was soon discovered that a large percentage of this wood was of inferior quality, and during stress testing individual spars broke at load factors of 1.1 or even a factor of 0.9. Furthermore, the glue used in the laminates was found to be of poor bonding strength, by which time the first 30 Ju 322s had already been partially assembled.

Once the Ju 322 V1 was completed it was placed on the take-off trolley for the purpose of conducting tank loading trials using the loading ramp. In doing so, the tank driver maneuvered so roughly that once the tank reached the tip of the ramp it dropped into the cargo hold with such force that the tank fell through the flooring. One the one hand, this event led to the cargo hold's flooring being reinforced considerably, resulting in a drastic empty weight increase for the Ju 322 and limiting its payload capacity to just 12 tons. On the other hand, a longer ramp was used for subsequent testing in order to reduce the pivot angle and prevent the onloading vehicle from tipping over. Later trials did not see a repeat of the accident during the first loading test.

Now the focus was on getting this oversized glider into its element, apparently a matter that up to that point had not been seriously considered. It was suddenly discovered that there was no suitable tow plane capable of lifting this monster into the air. The so-called "troika tow" had not been adequately tested at that point, and the Heinkel He 111 Z "*Zwilling*" did not yet exist. There was nothing left to do but try it with a Ju 90, which had already successfully towed the Me 321 "*Gigant*" aloft.

For testing the cargo glider, now given the appropriate covername of "*Mammut*" (Mammoth - with the covername "*Goliath*" also used in the beginning), a special detachment was set up within XI *Fliegerkorps* in early 1941 under the designation "*Merseburg*". By March of 1941 Special Detachment Merseburg had already taken acceptance of the first prototype, the Ju 322 V1 *Mammut*. For its maiden flight the Ju 322 was placed on an 8-wheeled takeoff trolley, which would remain on the ground once the cargo glider lifted off. A minimum ballast weight of four metric tons had to be carried onboard before the Ju 322 could be flown 'empty'. To get this colossus of a cargo plane into the air, the Junkers Ju 90 (*Werknummer* 0002) - at the time the most powerful four engine plane in Germany's inventory with its 1,250 hp American Twin Wasp radial engines - was harnessed ahead of the Ju 322. Yet even this machine was only suited for towing an unloaded Ju 322.

In a wise move, a 5 km corridor was cut through the forest at the edge of the airfield to clear more room for takeoff, and an overland guide was run in the direction of takeoff. For towing, a steel cable 16 mm thick and 120 m long was employed. After the team was in place and the enormous difference in size between the Ju 90 and the Ju 322 could no longer be overlooked, even the most stubborn optimists began having their doubts about the first takeoff attempt actually succeeding. An initial takeoff attempt on 6 March 1941 promptly ended in failure when the Ju 90 was unable to pull the 45 ton Ju 322 to liftoff speed and had to break off the run. With

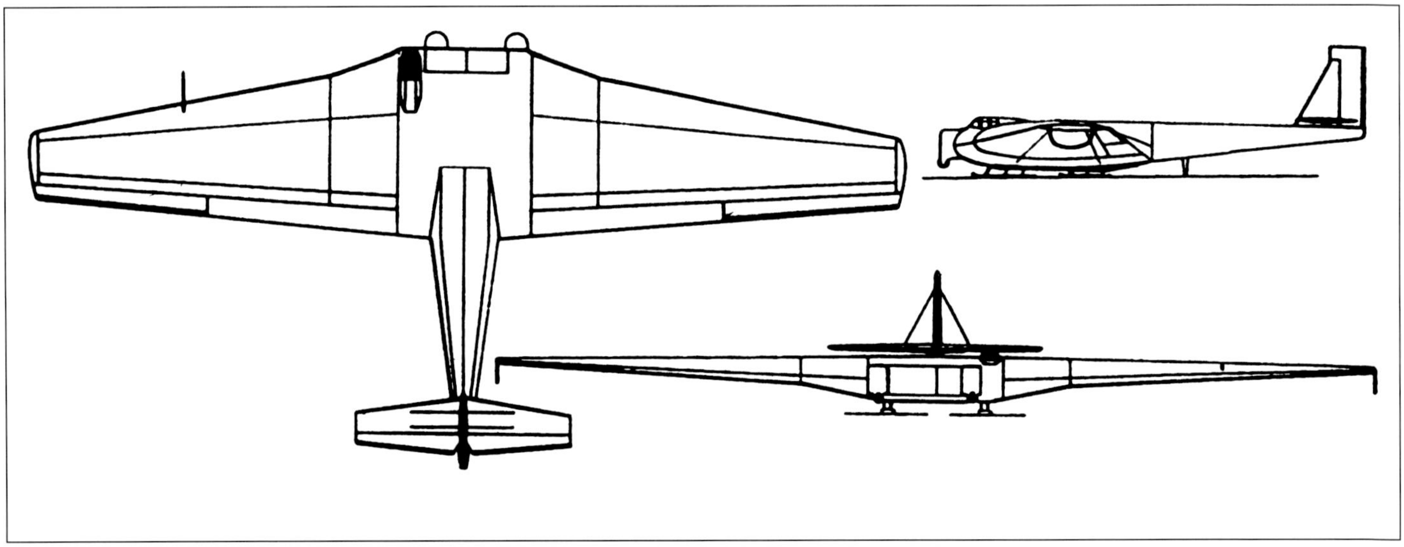

Reconstruction three-view of the actual layout of the Ju 322 V1, as it must have looked based on available photos.

The massive Junkers Ju 322 V1 at the airfield in Merseburg. To camouflage it against the prying eyes of reconnaissance aircraft, the extra large glider has been covered with innumerable camouflage netting sections. The Ju 322 has already been placed on its takeoff trolley in anticipation of its maiden flight. Despite the netting, it is possible to make out the two ballast containers next to the loading hatch. The pilot cockpit also juts out from the camouflage netting. In spite of the camouflage measures, the wooden behemoth did not remain hidden from British reconnaissance planes. In the background can be seen a few buildings at the Merseburg airfield.

considerable expedient optimism and a lot of sweat the Ju 322 was brought back to its takeoff point. Another attempt was hoping to be made, but this time the takeoff point was moved back to the very end of the runway in order to provide a little more run.

On 12 March everything was finally ready. With pilots Peter Hesselbach and Alfred Funke in the Ju 322 and Konrad Pernthaler with test engineer Anton Enders in the Ju 90 the team started out on its second attempt. Slowly, ponderously, the team began moving, despite the Ju 90's engines running at maximum power. The Ju 90 had already lifted off, while the Ju 322 stubbornly stuck to the ground - it simply wasn't able to reach the necessary speed for liftoff. Just before the end of the runway, when disaster seemed unavoidable, the Ju 322's pilots, seated side-by-side, forcibly yanked the glider into the air - taking the trolley along with it. From an altitude of a few meters the takeoff trolley finally separated, falling back to the ground and shattering to pieces. Freed from the trolley's tons of weight, the Ju 322 leaped higher. The Ju 322 swung to the side and bounced up and down. The crew of the Ju 90 made a futile attempt to accelerate and build up speed, and the Ju 322's crew tried to help by nosing the glider over to take load off the tow cable and give the Ju 90 room to accelerate. But this meant that they were going from the frying pan into the fire, for when the Ju 322 nosed over it increased its speed and closed the distance to the Ju 90 by a considerable margin, so that the Ju 322 had to be pulled upward again. The resulting lift created by the massive wings of the Ju 322, now flying much faster, had been vastly underestimated. The Ju 322 shot upward and yanked the tail of the towing Ju 90 so forcefully that the crew of the Ju 322 had to break the tow link to prevent the Ju 90 from crashing. With engines screaming the Ju 90 shot earthward and was able to pull out just before impacting the rapidly approaching ground. For its part, the Ju 322 went into a seemingly never-ending, long-distance glide. Fortunately there were a number of

The camouflage netting has been removed. The takeoff trolley on which the Ju 322 rests is now quite visible. In the event that a takeoff had to be aborted and the Ju 322 could land immediately, the aircraft was loaded onto the trolley with its landing skid already extended. One can see the brake shaft, used to help as a load balance, protruding from the lower fuselage. The wing join at the fuselage gives a good indication of the enormous profile thickness. On the upper wing can be seen the streamlined design of the pilot cockpit. All in all, this picture reveals that every three-view of the Ju 322 published up to this point has been erroneous with regard to the shape of the fuselage and empennage. Another interesting detail is that the photograph was apparently taken from a vehicle or airplane passing by.

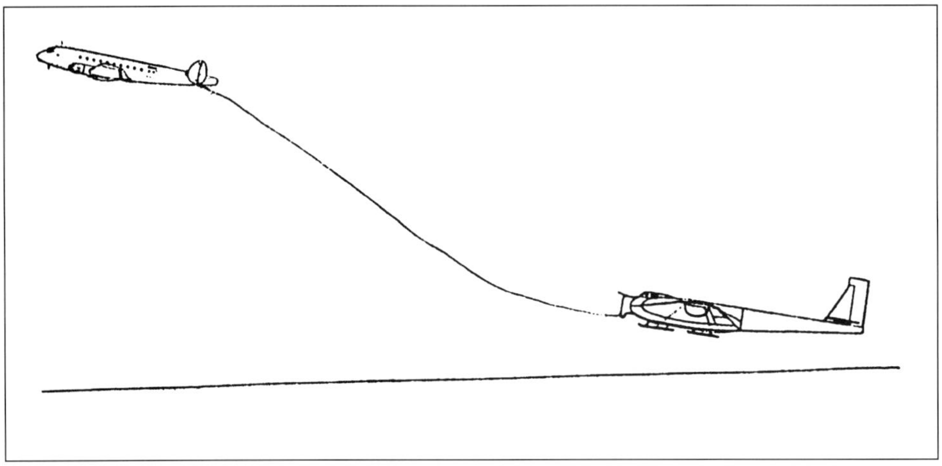

The sketch shows the situation immediately after liftoff. It still looks as though the flight might proceed more or less according to plan.

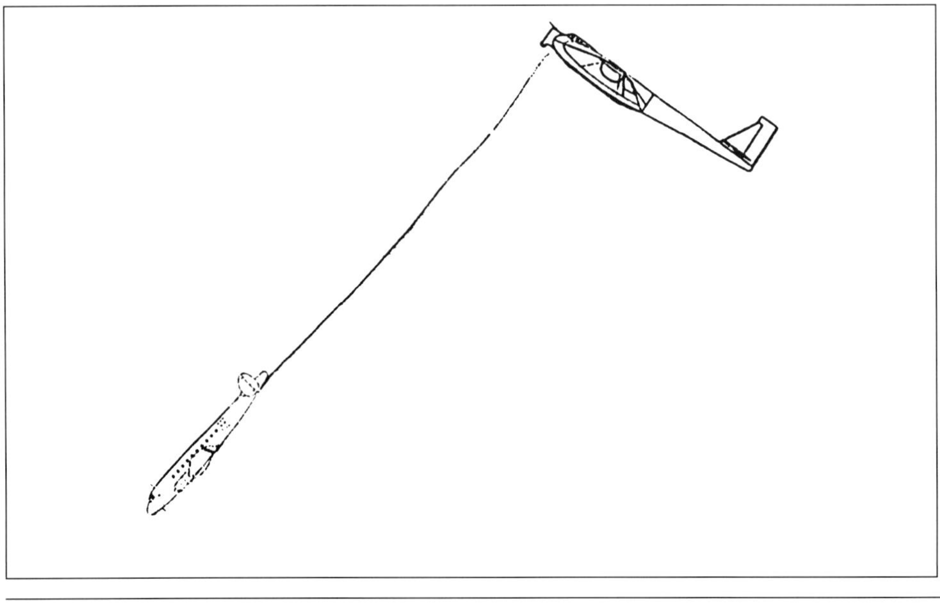

The situation has dramatically taken a turn for the worse. This must have been what the attitudes of the two aircraft were shortly before the tow cable separated.

fields stretching in front of the Ju 322's direction of flight, so that the flight path was completely free of obstacles. Helped by the ground effect, the Ju 322 simply did not want to set down; after all, a glide ratio of an improbable 1:50 had been attributed to the Ju 322, meaning that at an altitude of one kilometer it could travel the unbelievable distance of 50 kilometers with no wind, a feature that would make many a modern high performance glider turn green with envy. This unheard of glide capability was apparently the result of the wings thick, lift-generating profile stretching its entire length. This is known everywhere today as a laminar profile, but at the time it was unknown in Germany. Ultimately, the Ju 322 had to be forced to land by stalling it at ground level. With a rather hard bump it finally touched down and slid a few hundred meters on its skid before finally and abruptly coming to a stop in the soft earth of a field. *Dr.-Ing.* Fritz Buchwald, at the time a 20 year old young man, witnessed this event first-hand while serving with an anti-aircraft searchlight battery outside the village of Blösien near the airfield at Merseburg:

The Ju 90 tow plane vainly makes an attempt to gain altitude. The "*Mammut*", with its nose-high attitude, struggles above the shrubbery. The size difference compared with the Ju 90 is enormous.

In early 1941 I was stationed with an anti-aircraft searchlight unit near the village of Blösien west of Merseburg for protecting the industrial sites in the city. Our position was in an open field, and we had a good all-around view of the surrounding flat countryside. One day I became aware of the loud engines of an airplane coming from the direction of Merseburg and saw a Junkers Ju 90 approaching our position at low altitude and apparently at full throttle. This machine was towing a glider of a size I'd never before seen. The Ju 90 was the biggest plane I knew of at the time, but it looked absolutely tiny compared to the glider. I'd once been a glider pilot and watched the aircraft with great interest. The glider flew much higher than the Ju 90, which was vainly attempting to gain more altitude. Apparently the tail of the towplane was pulled sharply upward, and it seemed like a crash was imminent. Then the Ju 90 pointed earthward at an even steeper angle and raced downward when the glider detached itself. The Ju 90 pulled out just above the ground and flew off in a wide arc, but the falling tow cable cut our telephone line. The glider came towards us in a shallow, graceful glide and set down in a field about 200 meters beyond our position. A smooth landing, with hardly any damage caused. Since we assumed that this was a militarily important object, we immediately set up an initial cordon around the site, but I was unable to take a snapshot to remember this monster.

Over the next few days the entire area was sealed off by a special unit. Members of the airfield company unloaded a large quantity of ballast from the glider; rumors were that it totaled four tons - an unbelievable load for us. We assumed that the aircraft would be disassembled at the site and then transported away, but other plans were in the works. First a ramp was constructed up to a rail line that was slightly higher than the field. After about 14 days they tried to tow the aircraft off and pull it along the rail line using a tank. This attempt failed; the tank was unable to shift the aircraft even a centimeter from its position. It wasn't until they coupled a second tank in front of the glider that there was any kind of movement, and they towed the glider off in the direction of Merseburg. It wasn't until long after the war that I learned what kind of aircraft the gigantic glider was.

The photo opportunity that *Herr* Buchwald missed shows the Ju 322 V1 immediately after landing in a field. The wheels underneath the ballast containers served as castors, allowing a tractor to be hitched. The containers could be attached and removed using a crane and were filled with water as needed (maximum 2,000 liters/4 metric tons each).

This detailed report by *Herr* Buchwald clearly illustrates just how much effort it took to get the *Mammut* back on its 'feet' at the airfield following its successful landing outside the airbase perimeter. Once the Ju 322 had returned to the airbase at Merseburg and had been maneuvered onto its takeoff trolley, the team analyzed the events of this absolutely unsatisfactory towed takeoff. The empennage was enlarged and the attempt was made again, this time with a 'Troika tow' consisting of three Bf 110s, however. But even this second takeoff in late April 1941 was a failure. It ended in disaster. The Ju 322's lateral pendulum motion caused two Bf 110s to collide and crash. The Ju 322 made a forced landing and only the third Bf 110 survived unscathed.

It was decided that there was no sense in any further towed takeoff attempts, since there was simply no towplane available that could pull the empty *Mammut*, not to mention a fully laden Ju 322. With deep felt disappointment the entire project was halted; the prototypes already finished as well as the remaining 98 aircraft under various stages of construction were all sawn into pieces and scrapped. The piles of wood were burned off in innumerable wood combustion engines. 45 million Reichsmarks is what the *Mammut* project swallowed up. All documents were to have been destroyed - nothing was to be left to remind anyone of the disaster. Despite these instructions, a handful of documents and photos have survived.

What is left of this unfortunate design? Well, a flying machine was created that had the barely believable glide ration of 1:50. Yet what went wrong that this extremely positive quality of the Ju 322 wasn't exploited? This was certainly not your average glide ratio; it had a tactical advantage in that the Ju 322 could be released at an altitude of 4,000 meters above ground some 200 km from the landing area and then silently glide down to its target. The effects of the lift-generating wing, however, had been completely underestimated. The Ju 322 was thus able to lift itself into the air before the elevator barely had any effectiveness, and this was one reason why the Ju 322 swung sideways so violently on its one and only takeoff. At this point it behaved exactly like any other airplane that was flying on the verge of airflow separation; fortunately it did not show any tendency to stall over onto one wing. Its behavior normalized only after the glider detached itself, once enough speed could be built up, and this in turn was translated into a long distance glide - an absolutely praiseworthy quality that cannot be emphasized enough.

In order to overcome this disparity - generation of lift by the wing to the elevator effectiveness of the empennage - it would have been necessary to design the wing with sweepback while retaining all other components extant. This would have given the Ju 322 a much higher liftoff speed and, in so doing, ensured a greater elevator effectiveness, although naturally at the expense of the glide ratio. A further result of this measure would have been that the Ju 322 would have had a much higher in-flight speed, and, even more decisively, the lift energy being expended would have been reduced considerably with a swept wing than for a straight wing. You can therefore see that consideration of a swept wing for the Ju 322 would make sense. It's not hard to recognize that the modern jet-powered heavy lift transports have a relatively small empennage compared to quite a large wingspan - unlike those propeller-driven heavy lift transports still in service with unswept, straight wings, such as the Lockheed C-130 Hercules. The latter type has quite a large empennage to offset the problems described above. Furthermore, a heavy lift transport with swept wings is content carrying its payload with a relatively small wingspan. To put it in perspective, the transport capacity of the Galaxy would be inconceivable with a straight wing as found on the current transports of the Hercules or Transall class.

This is due to the fact that the swept wing justifies its use not only because of the potentially high speed it offers, but also because it provides the ability to carry extreme loads with a minimum of wingspan. There's no shortage of examples to prove this fact,

This photo shows the aircraft after Gert W. Heumann's 'beautification' efforts. However, this wasn't just restricted to touch-ups, but also included the addition of a few details not on the actual aircraft - as seen in the following detail shot.

The enlargement clearly shows the "after-market" application of machine guns on the supposed gun positions. The Ju 322 flew completely unarmed. The open canopy hatch also has 'magically' appeared in the wrong place. In the photo, the canopy hatch's actual position is blocked by the second ballast container.

This detail photo once again hints at the Ju 322's massive size. The extended brake shaft as load support is clearly visible.

such as the Bristol type 167 Brabazon. Designed in 1943 as a gigantic airliner with a wingspan of 70.10 meters, it also proved to be a gigantic mistake; like the Ju 322, a lack of available suitable powerplants prevented it from overcoming the planned takeoff weight of about 131 metric tons, making it entirely unsuited for a planned postwar London - New York service (first flight on 4 September 1949). The project by Howard Hughes, his Hughes H-4 Hercules (better known as the Spruce Goose), the largest aircraft built and flown to date, failed for similar reasons. With a span of 97.54 meters it was expected to be able to deal with a takeoff weight of 181 metric tons. By the time it took off on its first and only flight on 1 November 1947, it was obvious to the experts that even the Spruce Goose was underpowered for its intended role. A straight wing with a thick profile generates considerable lift, to be sure, but in exchange also generates an enormous amount of drag - a problem that has apparently been consistently underestimated. Be that as it may, the hapless story of the Ju 322 stands as an example of just how things should not be done. Furthermore, the flagrant lack of suitable tow aircraft made the development of the Ju 322 nothing more than a farcical adventure from the outset.

It is not well known that the license and patent department at Junkers, at the prompting of Messrs. Dr. Bator and Dr. Lemme-Roscher, ran into conflict with the Gothaer Waggonfabrik in the matter of the Ju 322. The background to this was the belief that a patent had been violated with regard to the Junkers patent DRP 688 757 for a "roll on, roll off transport aircraft" (pertains to the Ju 322), a feature that was believed to have been incorporated into the Gotha Go 242 cargo glider. A court case developed, calling for the Gothaer Waggonfabrik to pay Junkers a license fee for every Go 242 transport glider built (production cost set at 30,000 RM). Looking at drawings or pictures of the Go 242, it's hard to see the similarities. Unfortunately, we don't know whether the Junkers Company was successful in its prosecution of the case - perhaps a reader might be able to shed more light on this.

And on that note we will depart from this unfortunate subject in order to provide a summary of what has certainly been quite an interesting theme, the subject of swept wing projects.

This photo shows as Go 242 being towed by a Hanomag tractor. One must almost involuntarily ask where the similarity between this design and the Ju 322 lies.

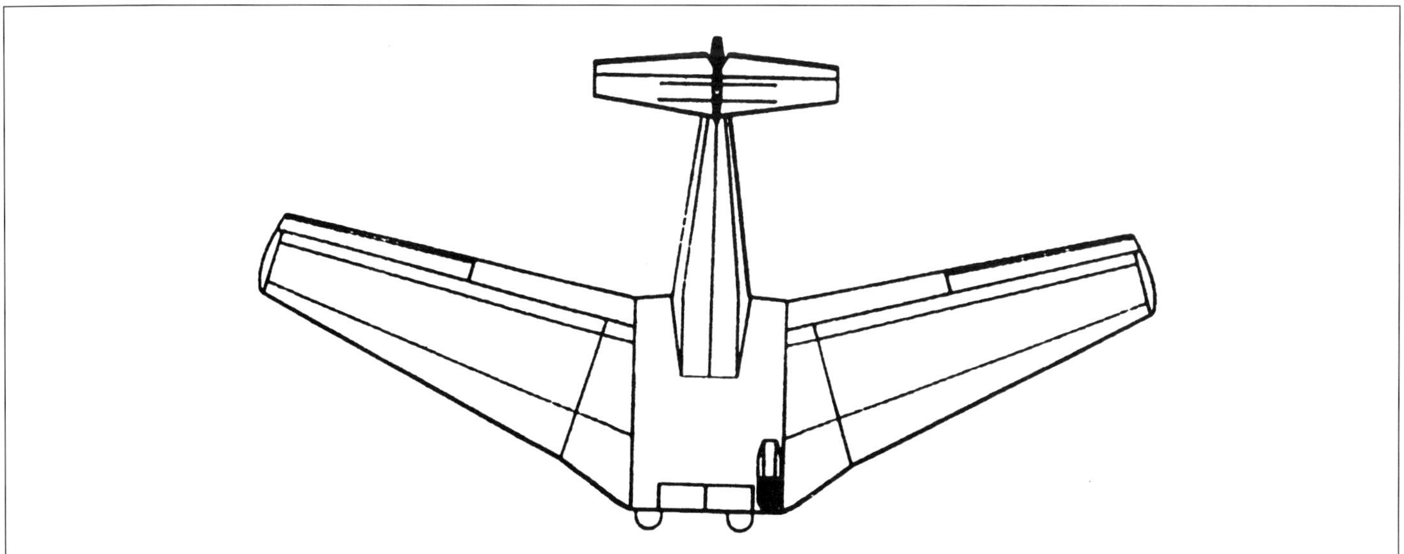

And to wrap up this section a sketch of the Ju 322 with its swept wings - it may look quite unconventional, but it would have been a most effective design (for what it's worth).

Epilogue and Credits

What do we have to show for all those hopeful developments? Of all the potential variants available, it was ultimately the wing with the positive sweepback that won the race. Whether a high performance fighter jet, a transport, ora passenger airliner - all turbine powered aircraft possess wings with a more or less positive sweep.

The negatively swept wing, on the other hand, was at first unable to make any inroads, since the main criticism was its lack of stiffness being seen as problematic. Only a very few postwar aircraft designs with negative sweepback are worth mentioning. Such as the HFB 320 Hansa, for example, a twin-jet low-wing design with a wingspan of 14.43 meters, a two man crew and the ability to carry seven to nine passengers at a maximum speed of 875 km/h, which was to have made its mark as a high-speed commuter and business executive aircraft. A few of these machines were ordered by the *Bundeswehr* as staff and liaison aircraft, and today there are probably one or two still in active service. The former Bavarian prime minister, Franz Josef Strauß, was the only German private pilot with a PPL-A license to fly the turbine-powered aircraft and landed in Moscow during the Cold War to attend a political meeting there.

In all, the HFB 320 was not a big success; only about 50 machines were ever sold. This was not due to its design or its performance, however. These were rated as being quite excellent. In-

The HFB 320 "*Hansa*" business and commuter jet. This photo shows the type's negative wing sweep to good effect.

stead, it was more due to poor marketing as well as a political climate that gave a German passenger plane little hope for breaking into the international market in the area of commuter and business traffic; simply put, the type lacked a good lobby. On the other hand, it should hardly come as a surprise to note that *Dipl.-Ing.* Hans Wocke was responsible for the HFB 320's design; after returning from the USSR, Wocke went to work for the Hamburger Flugzeugbau GmbH in Hamburg-Finkenwerder.

After the HFB 320 things grew quiet when it came to negatively swept wing designs. It wasn't until the 1980s that new life was breathed in to the idea with the American X-series line of experimental aircraft. With the X-29 and modern materials it was hoped that there would be advantages in the areas of maneuverability and turning radius, etc. But the concept remained with the experimentals for the time being. It wasn't until the appearance of the Russian Sukhoi S-37 *Berkut* that a new dimension in the future of negative sweepback seems to have opened up. Titanium and new types of bonding materials have given the wing the needed stiffness. However, only the future holds the answers as to whether or not the *Berkut's* extolled virtues compared to conventional fighters will have the impact needed for us to see other new designs shifting to negative sweepback.

Nevertheless, in conclusion it can be said that all those designs described in this book have found their modern successors in one form or another. The design features of the DFS 346 can be found in nearly all of today's swept-wing jets. The lines of the American B-2 betray its heritage from the Ho XIII/Ho XVIII. The Lippisch Li P 13, with its delta planform, has spawned quite a few progeny, e.g. the Concorde - once again making headlines. And a strong argument can be made for the Panavia MRCA Tornado's roots being traced back to the Messerschmitt P 1101. Even the 'flying wing', the Junkers Ju 322 *Mammut* celebrates its resurrection in the form of NASA's unmanned Helios high altitude flying wing, with its straight, stretched, 'load bearing' wing having a span of 75 meters and powered by 14 electro motors fed by batteries and solar cells, giving it an endurance of several days at altitudes of up to 30,000 meters. And, as mentioned earlier, the concept of negative sweepback has enjoyed a rebirth with the Russian Sukhoi S-37 *Berkut*, thus completing the circle.

Another word or two on the subject of justifying the swept wing is appropriate at this juncture. The swept wing is primarily seen as offering advantages in connection with high speeds, but it has another highly prized advantage that is rarely, if ever, discussed in any literature. With relatively small spans compared to straight

This view of the Sukhoi Su-37 "*Berkut*" clearly shows the unusual layout of this fighter, with its negatively swept wings combined with canards on the forward fuselage.

wings it allows incredible payloads to be carried, a fact that has been mentioned several times already and can be verified by the following table.

The table clearly shows that the largest aircraft of all time, the Hughes H-4, was just able to carry about one-fourth the payload of today's largest transport plane, the Antonov An-225 - which has a wingspan some 10 meters less than the H-4.

Aircraft Type	Wingspan	Payload Capacity
Antonov An-225	88.40 m	250 t
Antonov An-124	73.30 m	150 t
Lockheed C-5	67.88 m	119 t
Hughes H-4	97.54 m	59 t

When designing a successful aircraft it is necessary to balance wing sweep, profile and wingspan against each other with a view towards the intended role - high speed fighter, high performance heavy lift transport etc. -. In nearly all efforts, then, taking into account its peculiarities the swept wing is far superior to the straight wing. Only in a very limited number of areas, such as extremely slow flight, does the straight wing offer any advantages: one of the reasons for the swing-wing design, an attempt to marry the advantages of the swept wing with those of the straight wing.

There are only scattered fragments left of the former Junkers production facilities, such as a small part of the wind tunnel (see photo right). What has survived, however, is the memory of a highly interesting feat of technical engineering, born in a very difficult time for Germany.

This documentation would not be complete without recognition of the immense support given by many dedicated assistants and experts and eyewitness still alive today. To them I owe my gratitude for access to documents, pictures, journals and many other items. My thanks also for the many meaningful telephone conversations patiently borne and the responses to countless questions. In addition, I cannot forget the valuable photographs, indispensable to such a work, and I thank all those who more than once dug into long-forgotten archives and brought little-known material into the light of day.

And so my appreciation goes out to *Herr Ing. grad.* Walter Rebaum who, as an eyewitness, was able to contribute many previously unknown details regarding the history of the Junkers Ju 287, *Herr Dipl.-Ing.* Heinrich Theuer who, with his unique knowledge, was able to fill many gaps in the story, as well as *Herr* Günter Gauglitz whose personal experiences were able to shed entirely new light on the history of the DFS 346, and *Herr* Günter Schröter who experienced first-hand the test flights of the EF 131 as a flight engineer.

I owe a particular debt of gratitude to *Herr* Dr. Peter Korell, who permitted me to publish invaluable drawings from his files. And special thanks goes out to *Herr* Manfred Boehme for making available a wealth of documentation on the Ju 287.

I also owe my thanks to Messrs. *Dipl.-Ing.* Hans Holzer (*Deutsches Museum*), Manfred Griehl and Leo Schmitt, three gentlemen who repeatedly opened their vast archives to me, as well as to Hans-Peter Dabrowski and Peter F. Selinger who, among other things, made available photographs - many previously unpublished - on the Horten Ho XIII. And to Horst Thürling (senior editor of the former aviation journal AVIATIK) for his support, as well as to *Herr Ing. grad.* Horst Schmidt, who breathed new life into many barely recognizable sketches and blueprints. And I must also thank my daughter Jennifer, whose artistic talent was able to make many a scene more impactful.

I cannot forget expressing my appreciation to: Hartmut E. Sänger who, without much ado, made available much information on his parents, Professor Dr. Eugen Sänger and Dr. Irene Bredt, as well as Dr. Michael J. Neufeld (Smithsonian National Air and Space Museum), *Frau* Buchwald, Peter W. Cohausz, Albert Kleikamp, Major General Ruhlen, Georg Schlaug, Günter Sengfelder with his inimitable models as well as his expertise on the subject of aircraft undercarriages, *Oberstleutnant* Stephan, University of Keele, and last but not least the team at Foto-SB Niederschelden: *Herr* and *Frau* Holdinghausen, Renate Fischbach and Hella Knieps.

The pitiful remains of the former Junkers Works' wind tunnel, photographed by Martin Seidel in 1999 at Dessau.

Sources and Bibliography

A. Company records, test reports, minutes, war prize reports, journals, letters, etc.

01. From reminiscences and journals, *Dipl.-Ing.* Heinrich Theuer.

02. Directive No. 1477, Main *Führer* Hq, dated 03/27/1945

03. Memorandum for record on unlawful use of *Reichspatent* DRP 688 757, dated 06/06/1941.

04. Memorandum for record on unlawful use of *Reichspatent* DRP 688 757, dated 06/18/1941.

05. Memorandum for record on periscope installation for the Ju 287, dated 07/18/1944.

06. Establishment of bomber units, dated 07/07/1944.

07. Ju 287 construction description, dated 06/04/1944.

08. Meeting minutes from 02/18/1944, with *Gen.Feldm.* Milch.

09. Meeting minutes from 02/25/1944, with *Gen.Feldm.* Milch.

10. Meeting minutes from 05/26/1944, with *Gen.Feldm.* Milch.

11. Meeting minutes from 05/27/1944, Rechlin Test Center.

12. Meeting minutes from 06/23/1944, Fighter Staff meeting.

13. Meeting minutes from 07/01/1944, with *Reichsm.* Göring.

14. Meeting minutes from 07/03/1944, Fighter Staff meeting.

15. Meeting minutes from 07/06/1944, Fighter Staff meeting.

16. Meeting minutes from 07/18/1944, Rechlin Test Center.

17. Meeting minutes with the *Führer*, 09/21 - 09/23/1944.

18. Meeting minutes, dated 01/10/1945, Armaments Staff

19. Meeting minutes from 01/20/1945, Rechlin Test Center.

20. Letter from *SS-Standartenführer* Dr. med. Martin Brustmann, dated 06/07/1944

21. Letter from *SS-Standartenführer* Paul Baumert, dated 06/22/1944.

22. Letter from *SS-Standartenführer* Dr. Rudolf Brandt, dated 06/23/1944

23. Letter to SS Main Command Office - *Amt X*, dated 06/15/1944.

24. Memorandum on the He 343 and Ju 287 jet bomber types, dated 06/16/1944.

25. Various letters, *Ing. grad.* Walter Rebaum.

26. Ju 287 design and construction description, dated 09/09/1944.

27. War journal, from 08/01/1944 - end of war.

28. Directive TD 45, dated 08/16/1944.

29. Addendum to Heinkel memorandum from 06/16/1944 and Junkers memorandum from 07/10/1944.

30. Minutes re: license negotiations with Japan, dated 02/02/1945.

31. Minutes re: license negotiations with Japan, dated 03/27/1945.

32. RAE report no. F.A. 259/1 tech. note no. Aero. 1703, October 1945.

33. Compilation report, dated 09/25/1944.

34. Note to *SS-Hauptsturmführer* Dr. Kurt May, dated 06/16/1944.

35. *Rückstoßmistel* overview, spring 1945.

36. *Über die Flügelgeometrie von Nurflügelflugzeugen;* presentation by Dr. Reimar Horten, October 1981.

37. Junkers Company operations report for the month of September 1944.

38. *Luftwaffe* Command Staff presentation notes; *"Flugzeuge mit Strahl- und R-Triebwerken"*, dated 11/18/1943.

39. Weekly report for the period from 08/06 to 12/08/1944.

40. Company drawings: Horten Ho XIII a, overview drawings, dated 11/06/1944.

Junkers EF 94, type drafts, dated 12/03/1940

Junkers Ju 287, blueprint no. 8-28700-1038, dated 07/18/1944.

Junkers Ju 287 V3, main plans, dated 07/18/1944

Junkers Ju 287, various project sketches.

Junkers Ju 287, Ju 287 B nose and main undercarriage.

Messerschmitt Me P 1101, design drawings (sketch), dated 01/20/1945.

Various drawings in no particular order.

41. Research report Br. B. No. G 862/41: *Hochtemperatur-Strahltriebwerk*, Eugen Sänger

42. Research report No. 3509: *Über einen Lorinantrieb für Strahljäger,* Eugen Sänger/Irene Bredt.

B. Publications

43. Bartz, Karl, *Als der Himmel brannte* (Hannover 1955)

44. Butler, P.H., *War Prizes* (Leicester 1994)

45. Bukovski, Helmut & Griehl, Manfred, *Junkersflugzeuge* (Friedberg 1991)

46. Dabrowski, Hans-Peter, *Deutsche Nurflügel bis 1945* (Friedberg 1995)

47. Dressel, Joachim & Griehl, Manfred, *Heinkel He 177-277-274. Eine luftfahrtgeschichtliche Dokumentation* (Stuttgart 1989)

48. Dressel, Joachim & Griehl, Manfred, *Die deutschen Kampfflugzeuge im Einsatz* (Friedberg 1990)

49. Ebert, Hans J. *Willy Messerschmitt, Pionier der Luftfahrt und des Leichtbaues*, Die Deutsche Luftfahrt vol. 17 (Bonn 1992)

50. Georgii, Prof. Dr. Walter, *Forschen und Fliegen* (Tübingen 1954)

51. Green, William, *Warplanes of the Third Reich* (London 1970)

52. Griehl, Manfred, *Jetplanes of the Third Reich - The Secret Projects Volume One* (Sturbridge 1998)

53. Griehl, Manfred, *German Jets 1944-1945* (London 1988)

54. Hahn, Fritz, *Deutsche Geheimwaffen 1939-1945* (Heidenheim 1963)

55. Herwig, Dieter, *Geheimprojekte der Luftwaffe - Strategische Bomber 1935-1945* (Stuttgart 1998)

56. Horten Dr. Reimar & Selinger, Peter F., *Nurflügel* (Graz 1983)

57. Irving, David, *Die Tragödie der Deutschen Luftwaffe* (Frankfurt 1970)

58. Kens, Karl Heinz & Nowarra, Heinz J., *Die deutschen Flugzeuge 1933-1945* (Munich 1972)

59. Kössler, Karl & Ott, Günther, *Die großen Dessauer* (Planegg 1993)

60. Lippisch, Prof. Dr. Alexander, *Ein Dreieck fliegt* (Stuttgart 1976)

61. Lommel, Horst, *Geheimaktion Natter - Der erste bemante Raketenstart der Welt* (Stuttgart 1998)

62. Lommel, Horst, *Das bemannte Geschoss - Die Technikgeschichte* (Zweibrücken 2000)

63. Lommel, Horst, *Vom Höhenaufklärer bis zum Raumgleiter - Geheimprojekte der DFS* (Stuttgart 2000)

64. Möller, *Dr. Ing.* Eberhard, *Marine Geheimprojekte - Hellmuthh Walter und seine Entwicklungen* (Stuttgart 2000)

65. Neetzow, Klaus & Schlaug, Georg, *Deutsche Lastensegler 1938-1945 - Eine Chronik in Bildern* (Cologne 19xx)

66. Nowarra, Heinz J., *Die Deutsche Luftrüstung,* vols. 1-4, (Koblenz 1985)

67. Nowarra, Heinz J., *Junkers Großflugzeuge* (Stuttgart 1988)

68. Ransom, Stephen, *Zwischen Leipzig und der Mulde* (Lemwerder 1996)

69. Sengfelder, Günter, *Flugzeugfahrwerke* (Stuttgart 1979)

70. Sobolev, Dmitri Alexeevich, *Deutsche Spuren in der sowjetischen Luftfahrtgeschichte* (Hamburg 2000)

71. Thorwald, Jürgen, *Ernst Heinkel - Ein stürmisches Leben* (Munich 1976)

72. Wagner, Wolfgang, *Die ersten Strahlflugzeuge der Welt* (Koblenz 1989)

73. Ziegler, Mano, *Kampf um Mach 1* (Stuttgart 1965)

C. Other manuscripts, magazines, etc.

74. Aviatik, vol. 2-4, 1990

75. *Chronik Eugen Albert Sänger*, pp. 1-20

76. Der Flieger, vol. 3, 1950

77. Flight log of Hans J. Pancherz

78. Flugwelt, vol. 5-11, 1961

79. Monogram Close-Up 1: *Junkers Ju 287*, 1974

D. Picture Credits

Buchwald, Dr. Fritz (1), Cohausz, Peter W. (1), Dabrowski, Hans-Peter (7), *Deutsches Museum* (10), Griehl, Manfred (44), Kunze, Gottfried, via Manfred Griehl (2), Lommel, Horst (57), Nickel, Karl, via Peter F. Selinger (2), Rebaum, Walter (2), Ruhlen, Major General (1), Sänger, Hartmut E. (5), Schmitt, Leo (12), Schlaug, Georg (2), Selinger, Peter F. (8), Sengfelder Günter (15), Seidel, Martin (1), University of Keele (1), USAF Museum (4)

German Jet Aircraft 1939-1945

Hans-Peter Diedrich

Germany was one of the leading developers of jet propulsion during the World War II – in August 1939 the world's first jet aircraft, the Heinkel He 178, took to the air on its maiden flight. This new book examines all of the developments, production and aircraft types: He 280, Me 262, Ar 234, He 162, Ju 287, Ho IX, Me 328, P1101, Hs 132, DM 1, Ta 183 and others by such aircraft manufacturers as Heinkel, Junkers, Messerschmitt, and powerplant manufacturers BMW and Daimler-Benz.

Size: 8 1/2"x11" ■ over 80 b/w photographs ■ 112 pp.

ISBN: 0-7643-1230-8 ■ hard cover ■ $29.95

Arado Ar 234C An Illustrated History

David Myhra

The Arado Ar 234C was the world's first four turbojet-powered flying machine to be built in series. The Ar 234 was nothing less than Adolf Hitler's long awaited "Engländer Bomber." Powered by four BMW 003A-1 turbojet engines, with a combined thrust of 7,040 pounds, early test results indicated that it could reach speeds of over 550 mph even when loaded with tons of high explosive bombs and/or guided missiles. This complete illustrated history contains over 570 photographs and line drawings.

Size: 9" x 12" ■ over 570 photographs, line drawings ■ 192 pp.

ISBN: 0-7643-1182-4 ■ hard cover ■ 59.95